External Corrosion— Introduction to Chemistry and Control

AWWA MANUAL M27

First Edition

American Water Works Association

Copyright © 1987
American Water Works Association
6666 West Quincy Ave.
Denver, CO 80235

Printed in USA

ISBN 0-89867-366-6

Contents

Foreword, iv

Introduction, v

Chapter 1   Importance of Controlling External Corrosion 1
  Corrosion—Occurrence and Implications, 2
  Economics of Corrosion Control, 4
  Responsibility for Corrosion Control, 5

Chapter 2   Chemistry of Corrosion 7
  Basic Electrochemistry of Corrosion, 7
  Chemistry of Corrosion in Water Systems, 13

Chapter 3   Evaluating the Potential for Corrosion 20
  Effects of the Chemical Environment on Common Water Pipe Materials, 21
  Stray Currents, 31

Chapter 4   Preventing or Reducing Corrosion 35
  Coatings, 35
  Cathodic Protection, 37
  Polyethylene Encasement of Gray and Ductile Cast-Iron Pipe, 45
  Materials Selection, 46
  Trench Improvement, 46
  Protective Methods for Specific Pipe Materials, 47

Glossary, 49

References, 53

Index, 55

Foreword

This is the first edition of AWWA Manual *M27*, *External Corrosion—Introduction to Chemistry and Control*. Committee members involved in its development and approval included the following:

Robert E. Behnke, *Chairman*
W. Harry Smith, *Vice-Chairman*
Joe A. Willett, *Secretary*

G.S. Allen	D.R. Reedy
M.L. Bowden	R.C. Robinson
W.C. Creasman	S.R. Saylor
J.R. Easterly	M.J. Schiff
L.R. Keyser	T.F. Stroud
Donald Knibb	David Tippin
Gerald Mahon	Harry Uyeda
P.I. McGrath Jr.	J.L. Villalobos
J.H. Miller	C.F. Voyles
Franklyn Pogge	Donald Waters
Allan Poole	Roger Zeef

Introduction

For many years, frequent replacement of corroded pipelines and equipment was an unchallenged cost for utilities where aggressive soil conditions promoted extensive external corrosion. The amounts expended to maintain system integrity in many areas were significant, and to those obvious costs could be added the less quantifiable costs of the threats to public health and the deterioration of public fire protection resulting from corroded water lines that were overdue for replacement. Prevention of external corrosion was a little-understood art, with limited chance of success.

Today, corrosion prevention and control has progressed to the level of a science, allowing reliable predictions of corrosive conditions and, more importantly, effective prevention or mitigation of corrosion where economically appropriate. Nonetheless, concern over corrosion remains very much in order. The concern today, however, is not simply with the financing of system replacement; rather, it is with the more difficult issues of how best to install and protect a system to minimize corrosion, and with the trade-off in costs of corrosion protection versus the extended life of the pipelines and appurtenances.

Both the technology and the economics involved in controlling external corrosion are complex areas, requiring a logical and well-considered approach by utility managers, operators, and consulting engineers, all of whom must be familiar with local conditions and available options. Not all environments are corrosive, not all materials corrode, and there is no single answer to all corrosion problems. In any given situation, the corrective procedure selected must be the one most appropriate for the material and environment involved, and at the same time be economically feasible.

This book is addressed primarily to the professional water utility operator, whose objective it is to provide safe water to the public. The text is intended to give the reader an understanding of how and why corrosion happens, how the corrosion potential of an environment is evaluated, and how many of the proven measures for corrosion prevention and control operate. For readers needing a review of theoretical concepts of basic chemistry and basic electrical circuit theory, the relevant sections of *Basic Science Concepts and Applications** are recommended.

The general principles and examples presented in this book are not intended to replace the services of the corrosion-control engineering specialist. However, methodical application of the principles introduced—i.e., determining the cause of corrosion, analyzing its extent, and considering appropriate procedures for prevention or mitigation—will lay the foundation for an effective corrosion-control program that will benefit the public and the utility alike.

**Basic Science Concepts and Applications*, American Water Works Association, Denver, Colo. (1980, 1984).

AWWA MANUAL M27

Chapter 1

Importance of Controlling External Corrosion

Deterioration of pipelines, valves, pumps, and associated equipment due to external corrosion is an important concern for many water utilities. At one time, corrosion was accepted as inevitable in many soils, and extra thickness for metal piping was often specified to extend its useful life. Today, a variety of techniques are available to eliminate or significantly reduce external corrosion. Determining the need for such corrosion-control measures and selecting the most appropriate techniques are the primary topics of this book.

Before dealing with the more technical aspects of corrosion control, however, some discussion of the related economic and managerial issues is in order. This chapter presents a brief introduction to the science and terminology of corrosion, then discusses the issues of cost, economic return, and managerial responsibility for external corrosion-control programs.

After completing this chapter, the reader should be able to
- define corrosion;
- recognize certain environmental conditions and items of water supply equipment that are often associated with external corrosion problems;
- understand the cost and extent of the corrosion problem;
- recognize potential hazards to public health and safety that may result from corrosion;
- understand the basic economic questions that must be asked when selecting measures for corrosion control;
- recognize the responsibilities for a corrosion-control effort that must be assumed by various utility personnel.

CORROSION—OCCURRENCE AND IMPLICATIONS

Corrosion is generally defined as an electrochemical reaction that deteriorates a metal or an alloy. For the purposes of this book, corrosion also includes the dissolving of other water system materials through contact with water or soil.

Corrosion is a natural phenomenon. Metals are normally found in their stable, oxidized (corroded) form in nature. Iron ores, for example, are found as iron oxides. These oxides are chemically reduced in the refining process to produce useful metal, with the iron atoms in the elemental (unoxidized) form. In the presence of oxygen and water, or under certain soil and electrical conditions, refined iron tends to return to its more stable form, iron oxide (rust). Some waters and some soils are especially favorable to corrosion.

Potentially Corrosive Conditions

There are several conditions that increase the likelihood that corrosion will occur in a water utility system

- different metals or alloys in contact with each other and with a common media, such as water or soil, that may be conducive to corrosion;
- great variances in soil in contact with metal or alloys;
- contamination of soil with refuse, cinders, coal mine wastes, or salts;
- naturally occurring corrosive soil;
- chemical contamination of soil may deteriorate water utility system materials.

These conditions, discussed briefly in the following paragraphs, are treated in detail in Chapters 2-4 of this manual. Where such conditions occur, the water utility operator and engineer should be especially alert to the selection of materials and preventive measures that will minimize the effects of corrosion.

Dissimilar metals. Iron, copper, and many other metals are used in water system piping, valves, pumps, and other equipment. For each application, the manufacturer selects a metal with appropriate strength, machinability, cost, and other properties. There is no single ideal metal or alloy that could satisfy the many requirements of water system equipment.

Unfortunately, whenever two dissimilar metals are immersed in a common corrosive medium (soil or water) and placed in contact with each other, a condition called a bimetal couple is established, which may significantly increase the likelihood of corrosion. How extensive any corrosion will be depends on the characteristics of the corrosive medium and the metals involved. Figures 1-1 through 1-3 illustrate common uses of dissimilar metals and alloys in water systems. Each of these situations, and many others, is a potential location for corrosion.

Soil variances. The composition of soils usually varies from point to point, especially with changing soil depth. In many cases, a single metallic unit (pipe, well casing, valve, etc.) may be in contact with two or more completely different soil types, such as clay-sand, clay-gravel, or silt-clay. Whenever this situation exists, the likelihood of corrosion is increased. The severity of the corrosion will depend on the soils and the metal or alloy involved.

Naturally corrosive soils. There is great variety in the tendency of soils to promote corrosion. As a general rule, swamps, bogs, peat soils, and alkali soils are corrosive, and low, poorly drained soils are more likely to be corrosive than those in well-drained areas. The corrosivity (also called aggressiveness) of a given soil can be determined by sampling, testing, and/or analysis.

Environmental contamination. In many urban areas, the history of street surfacing may offer clues concerning potential corrosion of underground water system materials. Older streets often were surfaced with cinders, then later paved. Cinders are aggressive to most pipe and valve materials, and their presence is a warning that serious corrosion may occur.

CONTROLLING EXTERNAL CORROSION

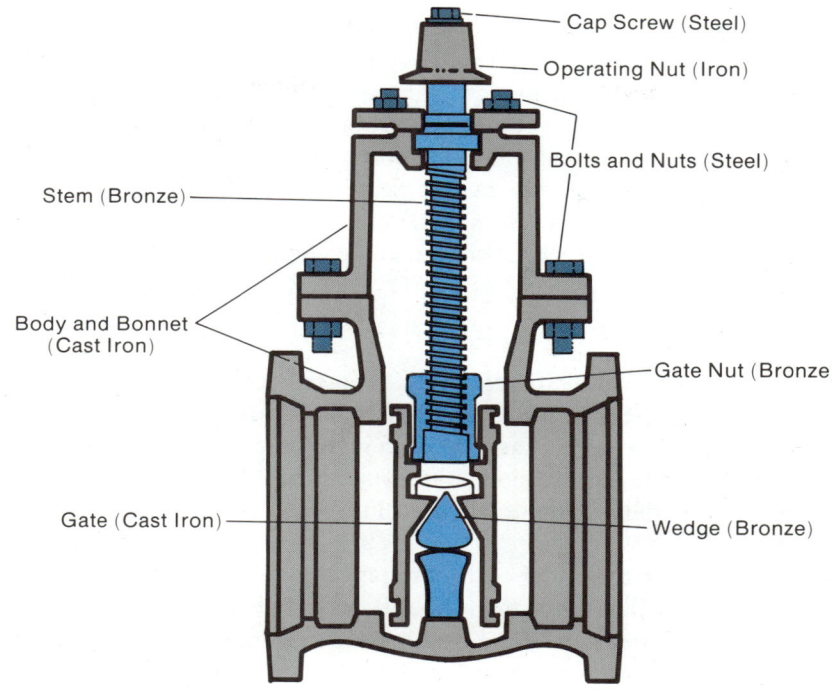

Figure 1-1   Metals Used in a Typical Gate Valve

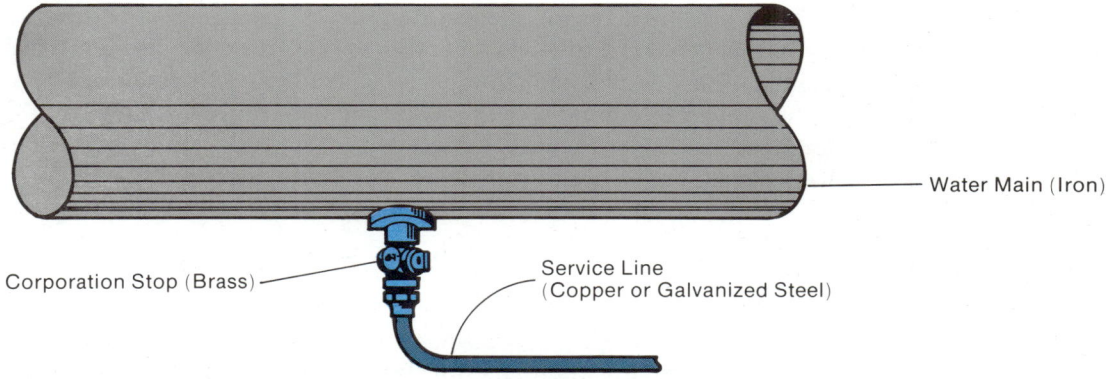

Figure 1-2   Metals Used at a Typical Water-Service-to-Main Connection

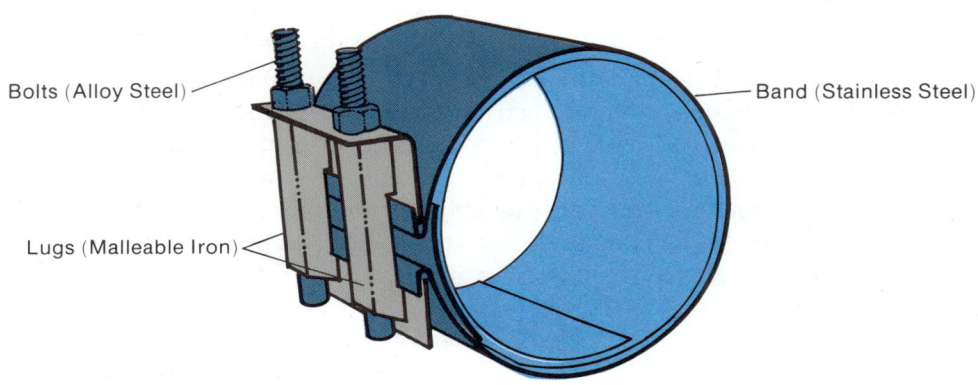

Figure 1-3   Metals Used in a Pipe-Repair Clamp

Chloride salts create corrosive soils. Steel reinforcement in concrete, iron, copper, brass, and many other materials in common use may be subject to attack if chlorides are concentrated in their environment. The current practice of heavy use of deicing salts on streets may be a source of future underground corrosion.

Sites where chemical contamination occurs, such as refuse dumps, landfills, and industrial waste disposal areas, may cause deterioration of water utility materials. Although the preferred policy is to avoid installation of underground materials in such exposure if possible, the potential for corrosion must be considered where alternate locations are not available.

Implications of Corrosion

The National Bureau of Standards has reported the total annual cost of corrosion in the United States to be on the order of $70 billion (for 1975), representing 4 percent of the Gross National Product at that time. Although there is no way of knowing how much of this cost is borne by the public water supply industry, the corrosive environment of many soils and the wide variety of materials used in a water system suggest that the percentage is significant.

In addition to the dollar costs of repair, replacement, labor, and equipment, there are even more important costs to the public as a result of corrosion. The health of water consumers is threatened whenever extensive corrosion breaches the sanitary integrity of the water system. The ever-present danger of backflow of contaminated liquid into the drinking water system is further increased when water pressure is interrupted to effect repairs on corroded wells, pumps, treatment equipment, pipes, valves, and services.

Another concern is that the safety of the public depends heavily on an adequate supply of water under pressure for fire control. There are hundreds of documented cases where low pressures and insufficient water resulted in the spread of small fires into disasters that caused injury, death, and needless destruction of property. Uncontrolled corrosion can be a major contributor to the problems of unreliable or inadequate fire-control systems.

It is evident that the ability to control corrosion in water utility systems can contribute greatly to dollar savings, public health protection, and safety of the public.

ECONOMICS OF CORROSION CONTROL

There are two primary considerations involved in any decision regarding corrosion control. The first, and most important, is the protection of public health and safety. The second is economics. Water utility systems are either publicly owned, requiring cost savings wherever possible, or investor-owned, in which case the owners expect and are entitled to maximum return on their investments.

A responsible water utility operator faced with decisions regarding what corrosion-control programs to implement must determine what actions will result in the greatest savings of money or the best return on money invested. The operator must ask which alternative is least expensive—to hold down initial costs and accept larger maintenance costs and shorter equipment life, to increase initial investment by specifying more corrosion-resistant material and reduce operating costs, or to invest money in corrosion-control procedures that will reduce maintenance and postpone replacement.

Return on investment. Probably the most common criterion used to compare the economic aspects of alternative solutions to the corrosion problem is return on investment. A general formula for return on investment is

$$\text{ROI} = 100 \left[\frac{(O_a + I_a/N_a) - (O_b + I_b/N_b)}{I_b - I_a} \right] \quad (1\text{-}1)$$

Where:

- ROI = return on investment
- O = annual cost of operation and maintenance
- I = installed cost
- N = anticipated life in years

(a and b refer to alternative installations).

The result of the formula, ROI, indicates what percentage of the additional initial investment for installation b will be returned each year as a savings over the long-term costs of installation a.

Economic evaluations are commonly the province of the design engineer and the utility management. Determining a reasonable estimate for the anticipated life of alternative installations requires considerable engineering expertise and experience. However, much of the data on which such evaluations depend is based on the experiences of operators with the existing system. As discussed in the next section, the responsibility for corrosion control is shared by many parties.

RESPONSIBILITY FOR CORROSION CONTROL

The responsibility for a logical water utility corrosion-control program lies fully with each of several involved persons: the consulting design engineers, materials producers, city officials or boards of directors (whichever applies), superintendents, and operators. Each must have sufficient basic knowledge on the subject of corrosion to make informed, appropriate decisions.

Although the consulting engineers need not be practicing corrosion specialists, they must possess enough background knowledge to enable them to recognize the warning signals and to know when it is necessary to call in a corrosion consultant.

City councils, village boards, water authority boards, and boards of directors cannot be expected to know the details of corrosion prevention and control; yet they may be legally responsible for accidents that result from lack of attention to corrosion or from improperly applied corrosion-control procedures. Therefore, they must take care to engage knowledgeable consultants and, if necessary, provide for training of the employees who design, build, operate, and maintain their water utility systems.

Producers of water utility materials have important responsibilities to the water utility industry. Because of the prevalence of soils and water exposures in water utility systems, material producers must ensure that information is furnished to the industry concerning any environmental factors that are corrosive to the materials, and that recommendations on best corrosion-prevention procedures for their materials are furnished to buyers.

Professional water utility superintendents and operators hold signal positions in the arena of corrosion control. Theirs is the responsibility for recognizing corrosion and its causes when it occurs, for recognizing potential corrosive situations before corrosion occurs or even before installations are made, for recognizing water quality problems that may stem from corrosion, and for knowing when corrosion is not a potential problem, thus saving money for their employers. Like the engineers, the superintendents and operators are not expected to be corrosion specialists, but they must be able to recognize the situation where the services of a consultant are indicated.

The responsibilities detailed in the preceding paragraphs are not to be taken lightly. The economic, public health, and fire-protection importance of corrosion prevention have already been detailed. Further, it is extremely important that no act in the corrosion-control effort create hazardous side effects. For example, the introduction of cathodic-protection electric current into the soil, without due consideration to stray-current damage to other underground structures, could result in accidents, property damage, and personal injury for which the water utility could be held legally responsible.

It is incumbent, therefore, on water utility operators, engineers, contractors, manufacturers, and officials to have a good, practical knowledge of external corrosion and corrosion prevention and control. This text is intended to form the basis for that knowledge.

Glossary

(Glossary terms are defined at the back of the book.)

Aggressive
Backflow
Bimetal couple
Cathodic protection
Corrosion
Corrosion control
Environment
Return on investment (ROI)

AWWA MANUAL M27

Chapter 2

Chemistry of Corrosion

Corrosion in water utilities affects a large number of materials, occurs under a wide range of environmental conditions, and can have greatly varying effects. Nonetheless, the physical mechanisms underlying corrosion are essentially unchanged from one situation to another. In this chapter, the basic chemical and electrical reactions of corrosion are explained, and several configurations in water utility systems are described where corrosion commonly occurs.

After completing this chapter, the reader should be able to
- explain the basic corrosion reactions;
- recognize an anode and a cathode on a corroded piece of equipment;
- explain the difference between a galvanic corrosion cell and an electrolytic corrosion cell;
- determine which metal will corrode in a bimetal couple;
- understand how polarization and passivation can retard corrosion;
- recognize several common types of corrosion and understand why they occur.

BASIC ELECTROCHEMISTRY OF CORROSION

Corrosion of metals and alloys is an electrochemical process; that is, a corrosion reaction involves both chemical reactions and the flow of electrons. There are two basic types of corrosion in a water system: galvanic corrosion and electrolytic corrosion. This section discusses the basic physical configurations and chemistry of both types.

Galvanic Corrosion

The galvanic corrosion process is identical to the reactions in an electrical battery, in which electrical current is generated by immersing two dissimilar metals, called electrodes, in a

8 EXTERNAL CORROSION

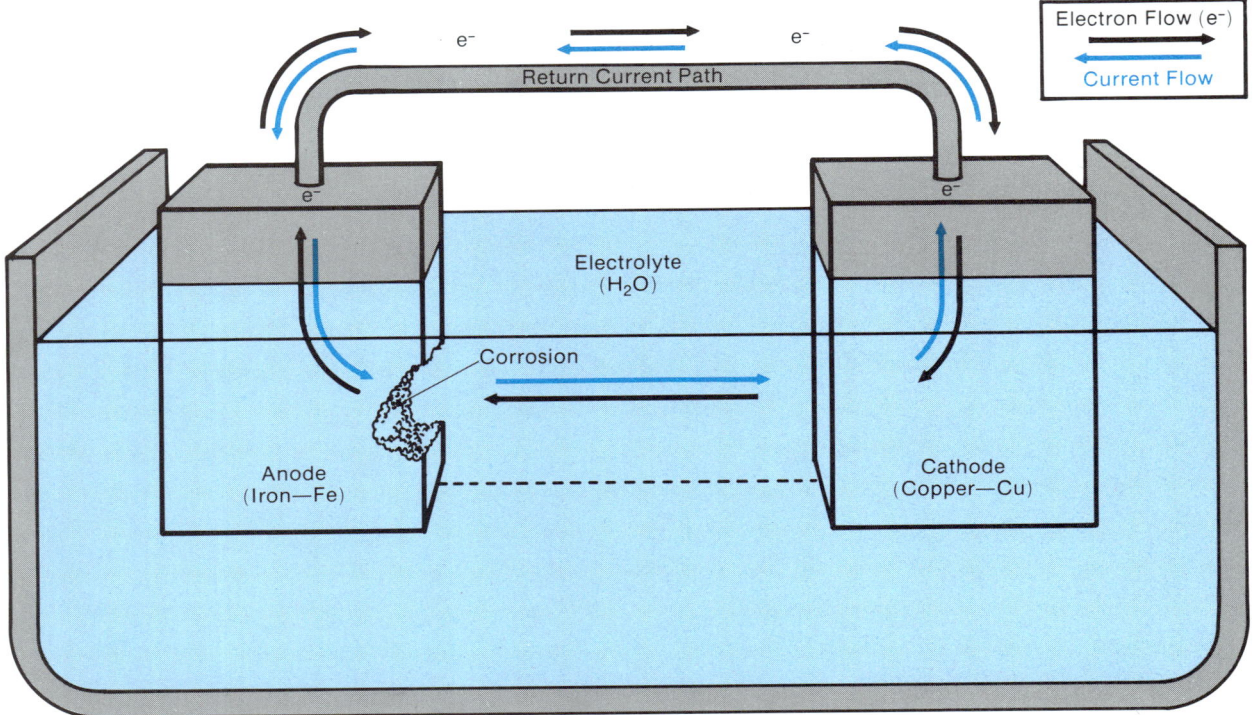

Figure 2-1 The Four Basic Elements of a Galvanic Corrosion Cell: Anode, Cathode, Electrolyte, and Return Current Path

chemical solution and connecting them with an external conducting wire, referred to as the return current path. Figure 2-1 shows the elements of a galvanic corrosion cell. Note the black arrows indicating the direction of electron movement and the blue arrows indicating current flow.

In the galvanic cell, corrosion takes place at the surface of the electrode where electrons are generated to travel through the electronic path. This electrode is called the *anode*. The conducting solution (water, soil, or some other chemical solution) is called the *electrolyte*. The electrode to which electrons flow is called the *cathode*. These four elements—the anode, cathode, electrolyte, and return current path—must exist before corrosion can occur.

Chemistry. The basic electrochemical reactions occurring in a galvanic corrosion cell are fairly simple. In the cell shown in Figure 2-2, the iron (Fe) anode on the left is corroding. Some of the iron atoms release electrons, which travel across the electronic path and enter the cathode. This loss of electrons changes the iron atoms from elemental iron (Fe^0) to ferrous iron (Fe^{++}) and then to ferric iron (Fe^{+++}), leaving them with a strong positive charge. Some of the molecules of the electrolyte, in this case pure water (H_2O), are naturally separated into hydrogen ions (H^+) with positive charges and hydroxyl ions (OH^-) with negative charges. The positively charged iron atoms are attracted to the negative OH^- ions. The attraction causes the iron atoms to leave the anode and enter the electrolyte, where they combine with OH^- ions. As atoms are lost, the metal surface of the anode deteriorates. This deterioration is corrosion. Note that the products of corrosion, $Fe(OH)_2$ and $Fe(OH)_3$, may accumulate on or near the corroded surface.

At the cathode, the negatively charged electrons arrive from the electronic path. The electrons are attracted to the positively charged H^+ ions in the electrolyte. The attraction causes the electrons to leave the cathode and combine with the H^+ ions, forming hydrogen gas (H_2). The gas may accumulate on the surface of the cathode. Note that the metal of the

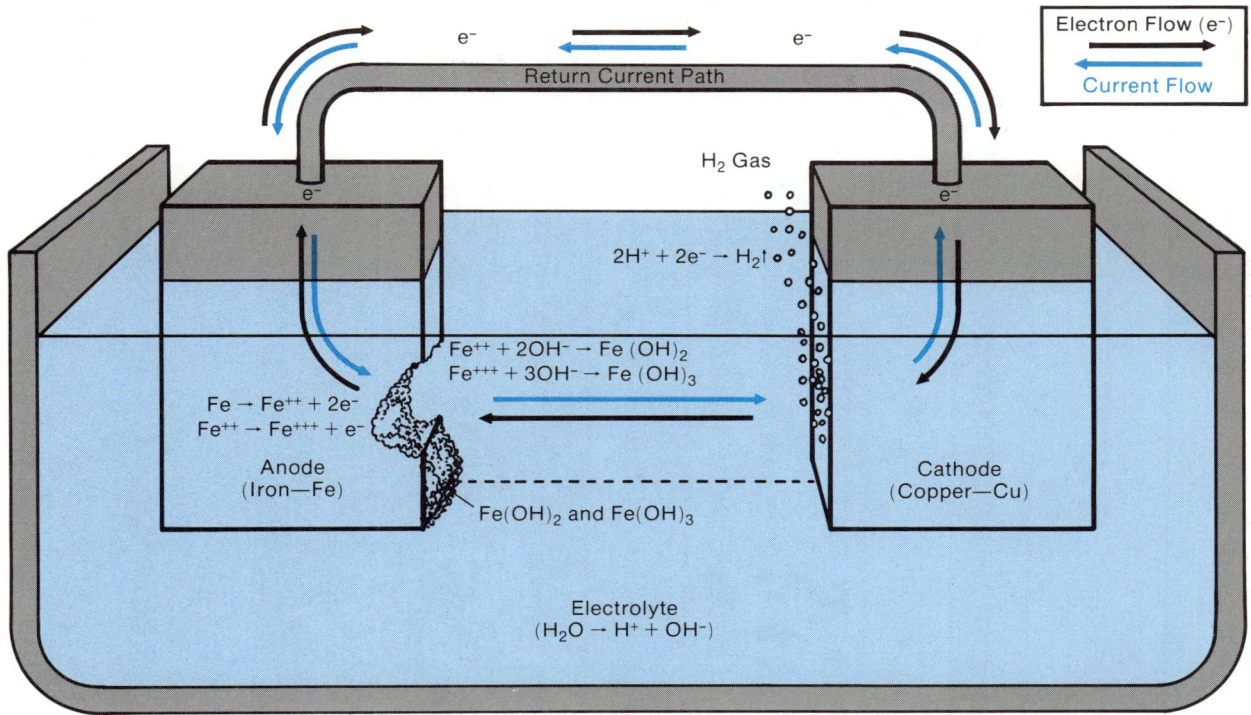

Figure 2-2 Chemical Reactions in a Typical Galvanic Corrosion Cell

cathode does not corrode. In fact, the reactions within the corrosion cell actually protect the cathode from corrosion.

Where different materials from the electrodes and electrolyte are used, the chemical reactions will be slightly different. For example, under field conditions, the water forming the electrolyte of the corrosion cell would often contain dissolved oxygen. If dissolved oxygen is added to the water in the corrosion cell just described, then the reaction at the cathode will combine the oxygen, water, and electrons to yield hydroxyl ions (OH^-) instead of hydrogen gas. Oxygen can react with the surface layer of hydrogen, thereby removing it as a depolarizing agent. Since the layer of hydrogen behaves as a resistor, its removal accelerates the cathodic reaction and corrosion activity. Changing the chemical composition of the electrolyte or the materials acting as the electrodes in a corrosion cell can cause great variations in the severity of the corrosion. However, no matter what materials are involved, it will always be the anode that corrodes and sends electrons into the electronic path, and it will always be the cathode that is protected from corrosion.

Nonuniform electrolytes. The type of galvanic cell just described has two different kinds of metal immersed in a single, uniform electrolyte. A second type of galvanic cell occurs where two pieces of the same kind of metal are immersed in a solid electrolyte of uneven composition, as shown in Figure 2-3. Because of the uneven character of the electrolyte, a corrosion cell can develop. One piece of metal will become the anode—it will corrode and feed electrons into the electronic path. The other piece of metal will be the cathode—it will be protected from corrosion and will feed electrons into the electrolyte.

As illustrated in Figure 2-4, the cell just described can be modified so that a single piece of metal acts as the anode, cathode, and return current path. Figure 2-4A shows the same configuration as Figure 2-3—two electrodes of the same metal in a nonuniform electrolyte. In Figure 2-4B, the two electrodes have been placed in direct contact with each other, eliminating the connecting wire. In Figure 2-4C, the two separate electrodes have been

10 EXTERNAL CORROSION

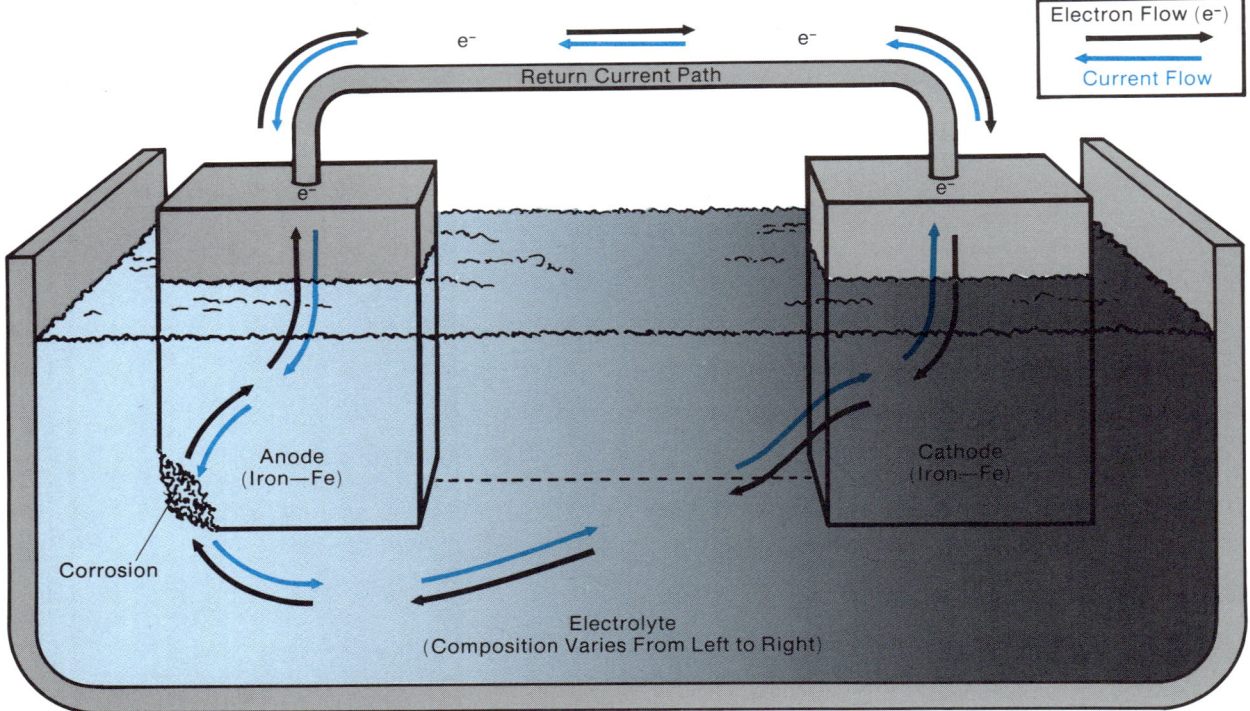

Figure 2-3 Galvanic Cell Formed With Nonuniform Electrolyte and Electrodes of a Single Metal

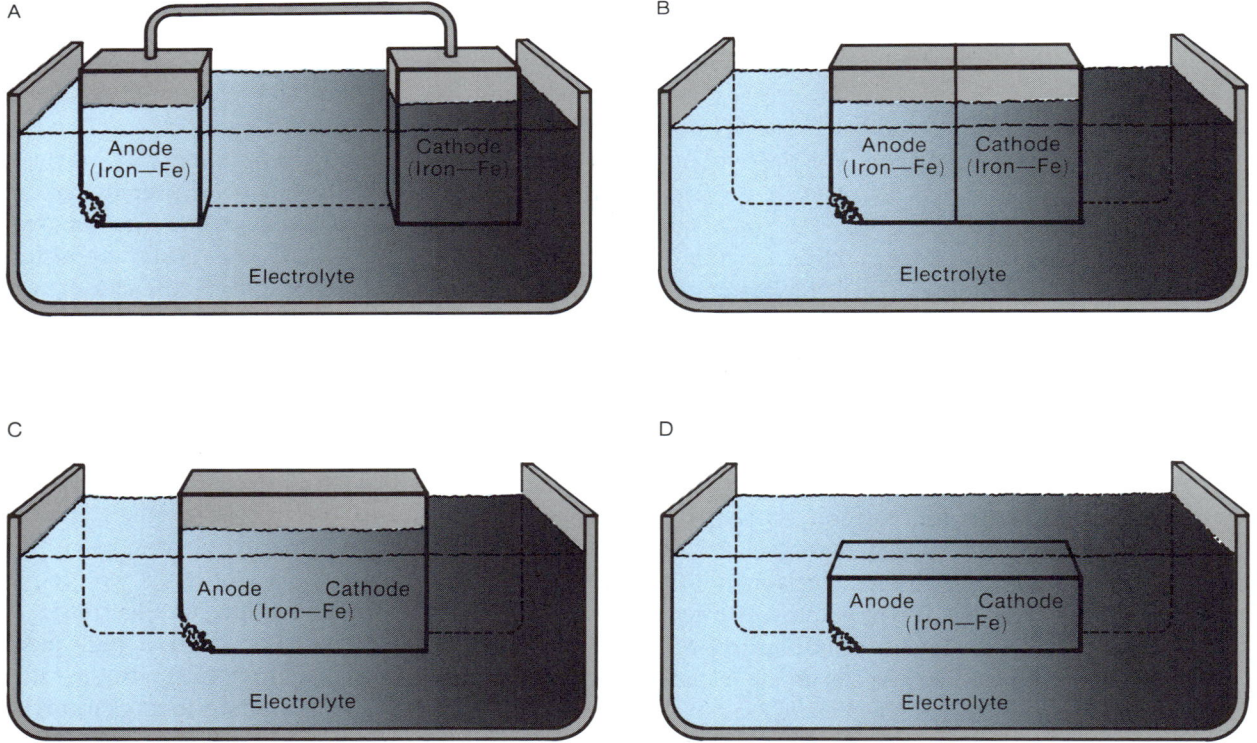

Figure 2-4 Creating a Galvanic Cell With a Single Piece of Metal in a Nonuniform Electrolyte

replaced by a single block of metal, which has one area acting as a cathode, another as an anode, with electrons flowing through the block as well. In Figure 2-4D, the block of metal, which could be a piece of pipe, has been totally buried in the uneven electrolyte, which could be soil. The effect remains the same as in the original configuration—the metal corrodes in the area that is acting as the anode.

Because soil is often nonuniform, there is always a possibility that such corrosion cells will develop around underground metal structures. Whether the cell causes significant damage depends on the strength of the soil as an electrolyte and the type of metal or alloy involved.

Current flow. Up to this point, the operation of the galvanic corrosion cell has been described in terms of the movement of electrons. In practice, corrosion cell chemistry is often discussed in terms of electrical current flow, also called conventional current flow. Because of a historical misunderstanding about the nature of electricity, this current flow is considered to move in the direction opposite to the movement of the electrons. Thus, where electrons leave iron atoms in the anode to travel into the electronic path, the conventional current flow is said to go the other way, leaving the surface of the anode and entering the electrolyte. At the cathode, where electrons enter the electrolyte, the conventional current flows from the electrolyte into the cathode.

Restated in terms of conventional current flow, the basic galvanic corrosion reaction is as follows (Figure 2-5): electrical current is generated by immersing dissimilar metals or alloys in a uniform electrolyte, or by immersing a single metal or alloy in a nonuniform electrolyte. The current travels from the anode through the electrolyte to the cathode and returns through the connecting current path. At the location of current discharge from the anode into the electrolyte, there is loss of anode metal. This loss of metal is corrosion. Corrosion control is the process of reducing, eliminating, or reversing the current flow, thus reducing or eliminating the corrosion.

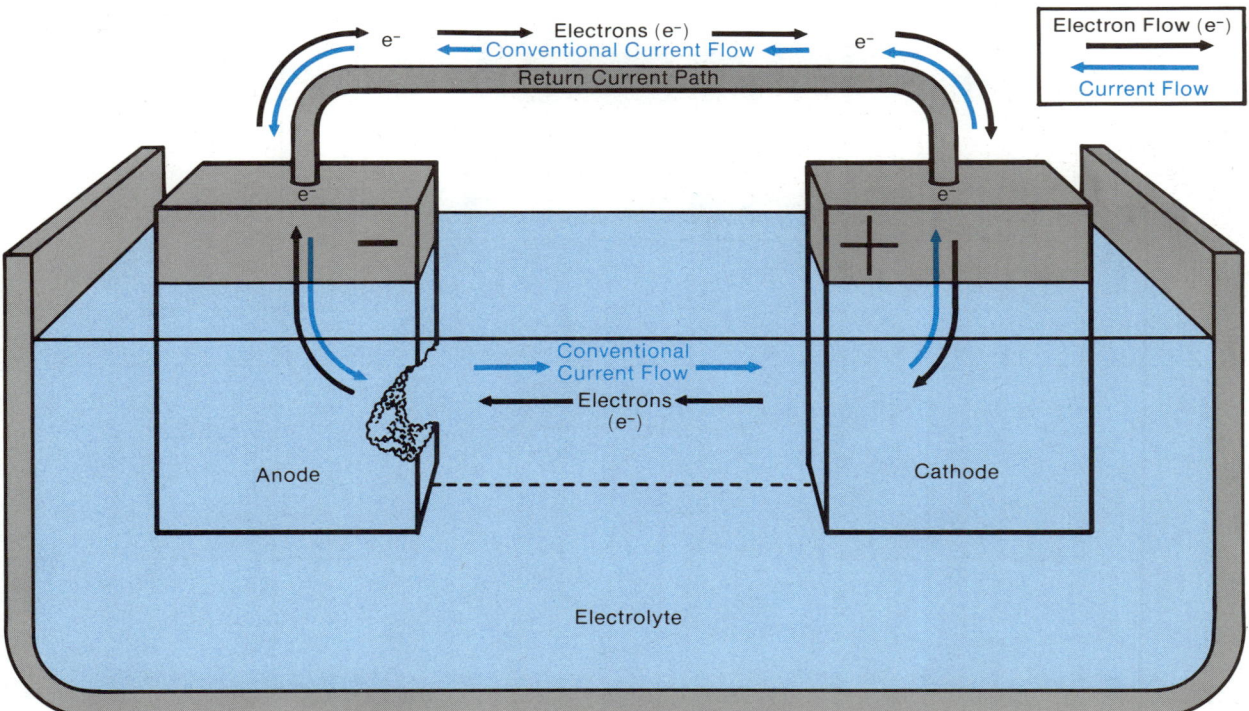

Figure 2-5 Contrasting Conventional Current Flow With Electron Movement in a Galvanic Corrosion Cell

Electrolytic Corrosion

The galvanic corrosion cell discussed in the previous section requires an anode, a cathode, an electrolyte, and a return current path. The reactions in the cell generate an electrical current. The configuration of an electrolytic cell is similar to a galvanic cell, but the electrolytic cell does not generate an electrical current. Instead, the corrosion reaction is driven by a direct-current source originating outside the cell. The end result, however, is the same—corrosion of the anode.

Figure 2-6 shows an electrolytic corrosion cell. The four basic elements of the galvanic cell are still necessary—the anode, cathode, electrolyte, and return current path. In addition, an outside direct-current source, located within the return current path, must exist to drive the reaction. As current is forced through the corrosion cell by the outside current source, corrosion occurs at the anode.

When electrolytic corrosion develops accidentally in metals exposed to soil or water, it can be severely damaging. There are several situations where the external direct current needed to drive electrolytic corrosion may occur in the area of a water distribution system. Direct-current-driven train or subway systems can be a source of stray current if the rails, the system's intended return current path, are not completely insulated from the soil (Figure 2-7). If underground pipe or other metallic equipment picks up a portion of this current, it will corrode at the point where the current ultimately leaves to rejoin the original circuit. Other sources of stray direct current are electric welding equipment grounded to underground utilities and stray direct current from nearby cathodic protection systems.

Alternating current does not have the same effect on metal at the point of discharge as direct current and generally is not a cause of corrosion. In rare cases, a small portion of alternating current traveling from an underground source can become rectified to direct current, which can cause corrosion.

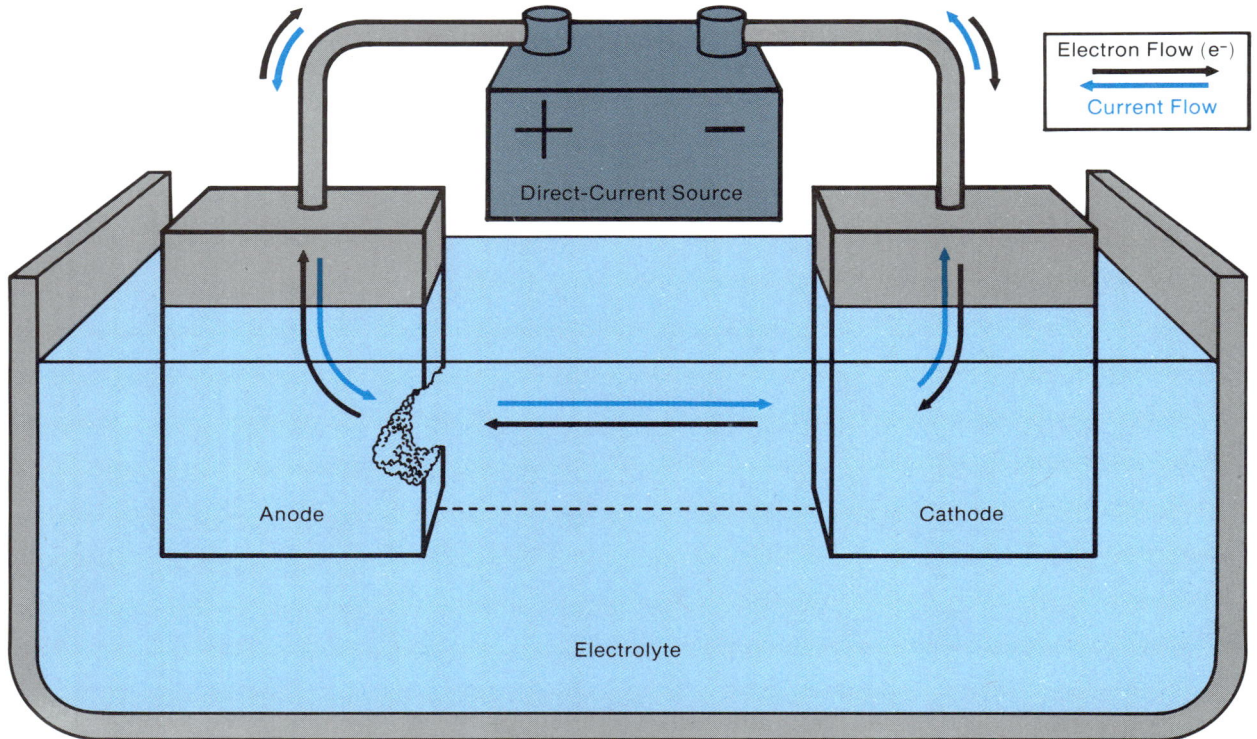

Figure 2-6 A Typical Electrolytic Corrosion Cell

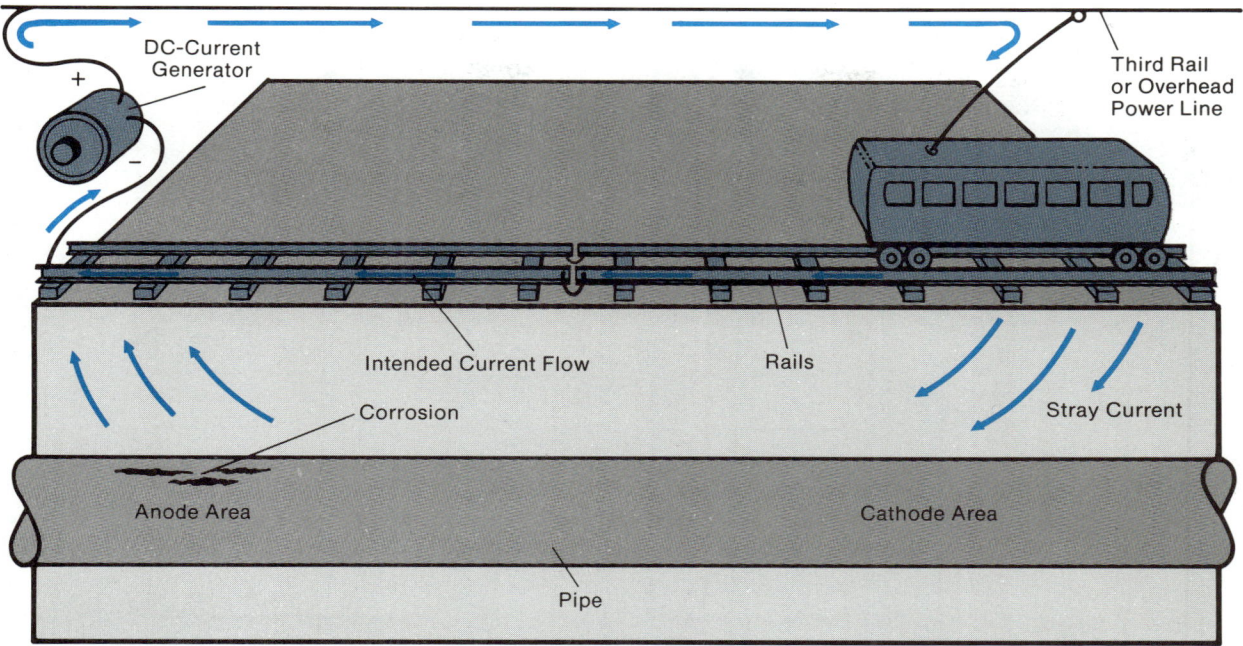

Figure 2-7 Direct-Current Transportation Systems as a Source of Current Causing Electrolytic Corrosion

CHEMISTRY OF CORROSION IN WATER SYSTEMS

All corrosion can ultimately be explained in terms of the principles discussed in the first part of this chapter. However, the nature and severity of corrosion found at various points within a water system will vary greatly depending on the materials involved, the electrolyte, the physical configuration of the cell, and the environment. This section discusses the calculation of corrosion rates and considers the effects of different materials that may be involved in a galvanic cell. It also explains two physical/chemical processes that may reduce the severity of corrosion, then briefly describes a number of common corrosion configurations found in water systems.

Calculating Corrosion Rate

Corrosion reactions happen in accordance with well-understood physical laws. By determining the conditions surrounding a corrosion cell, the theoretical rate of corrosion can be precisely predicted. In the field, changing conditions will vary the calculated rate somewhat, but the predictions are still an important tool in determining the need for corrosion-control measures.

Cell voltage. The chemical reactions occurring in a galvanic corrosion cell force electrons through the electronic path. When the return current path is disconnected, the force—the potential for current flow—still exists, exhibited as the difference between the electrical charges of the anode and the cathode. As in any electrical circuit, this electrical force is known as potential, or voltage. The voltage across the electrodes of a galvanic cell that is not connected to an electronic path is called the cell voltage, cell potential, or voltage differential. It can be measured by connecting the leads of a voltmeter to the two electrodes, as shown in Figure 2-8.

The chemical reactions within a galvanic corrosion cell will vary depending on the materials making up the cell, and the cell voltage will vary accordingly. For corrosion cells

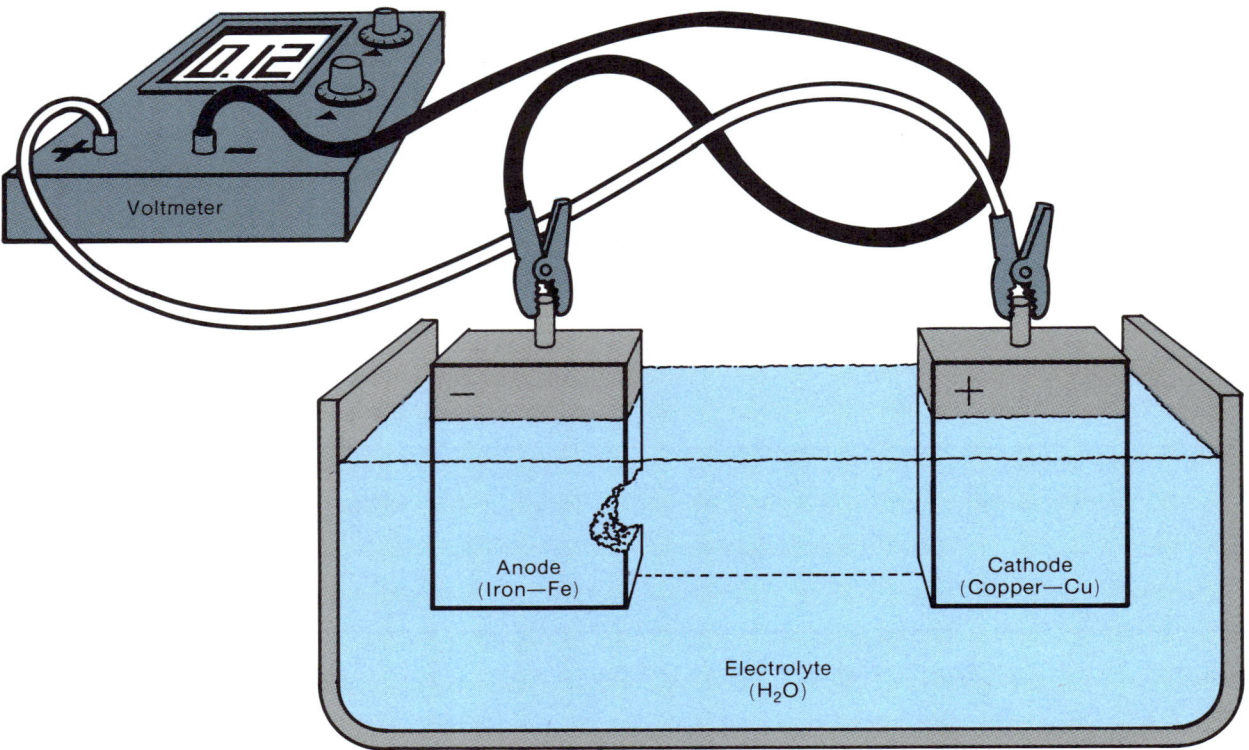

Figure 2-8　Measuring Cell Voltage With a Voltmeter

created as part of a battery designed to generate electrical current, the potential can be several volts. For corrosion cells commonly occurring in water utility piping systems, the voltage will usually be less than 1 V, and it is commonly measured in millivolts (1000 mV = 1 V). The cell voltage is the driving force that pushes the electrons through the electronic path—the greater the voltage, the greater the current, and the more rapid the rate of corrosion.

Ohm's and Faraday's laws.　The effects discussed in the previous paragraph can be expressed mathematically. Ohm's law states that the current flow through the electronic path will be equal to the voltage applied (the cell potential) divided by the resistance of the circuit. Thus,

$$I = \frac{E}{R} \qquad (2\text{-}1)$$

Where:

I = current flow (A)
E = potential (V)
R = resistance (Ω).

According to Faraday's law, the rate at which an anode of a given metal corrodes varies precisely with the magnitude of electrical current flow. The corrosion rate for iron, for example, is 20 lb (9.1 kg) per ampere of current per year. Using Ohm's and Faraday's laws, the rate of current flow and the rate of corrosion can be precisely calculated wherever the cell potential and the resistance of the return current path are known.

Galvanic Corrosion Materials

Galvanic corrosion cells can occur in many of the combinations of metals, alloys, and electrolytes found in all stages of potable water production and distribution. In considering corrosion cells caused by differential metal combinations (known as galvanic couples or bimetal couples), it is essential to know which of the two metals involved will act as the anode and, therefore, corrode. It is also important to know how great the potential for corrosion will be for any two metals. The galvanic series for a given electrolyte supplies the answers to these questions.

The galvanic series. Table 2-1 shows the galvanic series for seawater. When any two of the metals shown in the series are combined with the electrolyte and a return current path to form a galvanic cell, the metal nearer the top of the series will be the anode and corrode; the metal nearer the bottom will be the cathode and be protected. The metals on the cathodic end of the series are said to be more noble than those on the anodic end.

The farther apart two metals are in the series, the greater the cell voltage they will produce across the return current path when immersed in the electrolyte. The cell voltage and the resistance of the return current path determine how rapidly current will flow in the circuit and, therefore, how rapidly corrosion of the anode will occur. Thus, metals that are widely separated in the galvanic series will produce corrosion cells that show a high rate of corrosion, even when the electrical resistance of the return current path is high. Metals that

Table 2-1 Galvanic Series of Selected Metals and Alloys (in Seawater*)

Anodic, Active (Read Down)†

Magnesium
Magnesium alloys
Zinc
Aluminum 52SH
Aluminum 4S
Aluminum 3S
Aluminum 2S
Aluminum 53S-T
Alclad
Cadmium
Aluminum 17S-T
Aluminum 24S-T
Mild steel
Wrought iron
Gray and ductile cast iron
Ni-resist
13% Cr stainless steel, type 410 (active)
50-50 lead–tin solder
18-8 stainless steel, type 304 (active)
18-8, 3% Mo stainless steel, type 316 (active)
Lead
Tin
Muntz metal

Manganese bronze
Naval brass
Nickel (active)
Inconel—76% Ni, 16% Cr, 7% Fe (active)
Yellow brass
Aluminum bronze
Red brass
Copper
Silicon bronze
Ambrac—5% Zn, 20% Ni, 75% Cu
70% Cu, 30% Ni
Comp. G bronze—88% Cu, 2% Zn, 10% Sn
Comp. M bronze—88% Cu, 4% Zn, 6.5% Sn, 1.5% Pb
Nickel (passive)
Inconel—75% Ni, 16% Cr, 9% Fe (passive)
Monel—70% Ni, 30% Cu
18-8 stainless steel, type 304 (passive)
18-8, 3% Mo stainless steel, type 316 (passive)
Titanium
Silver
Graphite
Gold
Platinum

Cathodic, Noble (Read Up)†

*There is a specific galvanic series for each environment. The relative positions of the various metals and alloys may vary slightly from environment to environment.

†In a galvanic cell of two dissimilar metals, the more active metal will perform as the anode and be corroded, while the more noble metal will perform as the cathode and be protected.

are close together in the series will produce cells where little or no corrosion occurs, since the voltage produced will seldom be sufficient to force a significant current through the resistance of the return current path.

In the series shown in Table 2-1, the cell exhibiting the greatest corrosion would be formed with an anode of magnesium and a cathode of platinum, immersed in seawater and connected through a return current path. The galvanic series shown in Table 2-1 is accurate where the electrolyte is seawater. There would be slight variations in the arrangement of the metals in a galvanic series for fresh water or moist soil, and the magnitude of the cell voltage generated with electrodes of any two metals would vary considerably.

Galvanic materials in water systems. In most water systems, equipment can be found containing parts made of steel, gray and ductile cast iron, stainless steel, lead, brasses, copper, and bronzes. Many of these materials are exposed only to treated water, and although these components may be subject to internal corrosion within the electrolyte of the treated water, that subject is beyond the scope of this book. However, combinations of metals exposed externally do occur, notably where services, valves, or pumps are connected to mains.

For example, a brass curb stop connected to a wrought-iron service can be a source of considerable corrosion, as implied by the distance between brass and wrought iron in the galvanic series. A typical large valve may utilize several metals and alloys in its construction, with two alloys commonly used on the outside exposure: steel for nuts and bolts and cast iron for the valve body. It would appear that the two alloys are too close together in the galvanic series for corrosion to occur; however, the tendency for steel to act as an anode in a cell with cast iron is greatly magnified by the difference in surface areas between the small bolts and nuts and the much larger cast-iron valve body and bonnet. The same effect is notable in gray or ductile cast-iron mechanical-joint bolts and nuts, as well as in many other configurations within a water utility system where a small surface area of relatively anodic metal is found in contact with a large surface area of relatively cathodic metal. One good method of counteracting this differential effect of surface area is to provide greater nobility in the smaller metal unit. For example, ductile-iron pipe is furnished with alloy steel coupling bolts containing copper, nickel, and chromium. This effectively prevents significant corrosion reaction even in fairly conductive electrolytes.

The most severe galvanic corrosion occurs where two metals are installed that are widely separated in the galvanic series. However, even a single piece of metal equipment will have slight variations in the composition of its metal surface from one point to another. In a sufficiently strong electrolyte, these variations can produce microscopic galvanic corrosion cells, which can eventually grow and cause major deterioration of the metal.

Polarization and Passivation

Two different effects that may reduce the rate of corrosion are polarization and passivation. Where appropriate, a design engineer may select materials that encourage these effects.

Polarization. As the products of the corrosion reaction build up at the anode and cathode, the voltage difference between the two tends to diminish. This is called *polarization*. The effect is likely to be greater at the cathode. For example, hydrogen gas can literally blanket the cathode surface, tending to retard the flow of current. The hydrogen layer may be partially removed as hydrogen gas combines with oxygen gas, forming water. This is more pronounced on a comparatively large cathode because the hydrogen is spread thinly over a large surface area. Hence, it is safer to have a large anode and small cathode than the reverse.

Even so, the anode can become polarized by a layer of oxide film or other compound that has the effect of forming a barrier between the metal and its environment. Some metals polarize easier than others and, therefore, metal selection can be an important part of engineering.

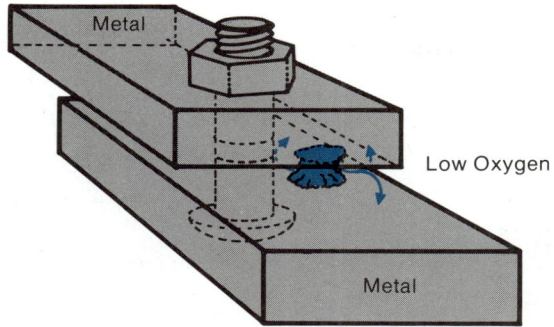

Figure 2-9 Concentration Cell (Crevice) Corrosion

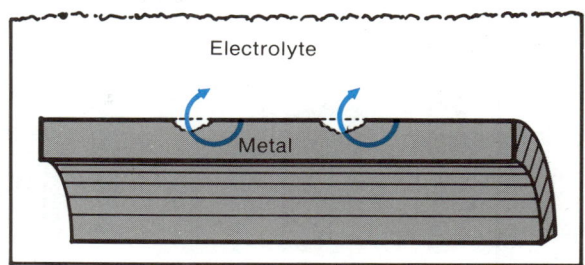

Figure 2-10 Pitting Corrosion

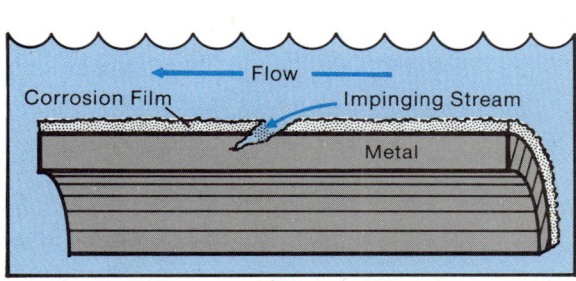

Figure 2-11 Impingement Corrosion

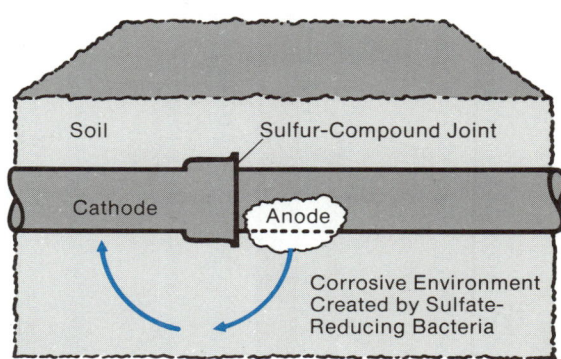

Figure 2-12 Bacterial Corrosion

Passivation. Passivation is a condition in which a given metal behaves in the manner of a more noble metal than is indicated by its position in the galvanic series. It is a condition of electrochemical activity where the initial corrosion products of certain metals provide protection of the base metal.

Specific Types of Corrosion in Water Systems

A number of situations are commonly found in water utility systems where a combination of environmental factors, physical configurations, and materials have accelerated corrosion. The next paragraphs describe some of those conditions and their effects.

Concentration cells. Concentration cells, Figure 2-9, are localized occurrences of corrosion that often develop at crevices in metal units due to the uneven diffusion of oxygen. Oxygen content inside the crevice may be very low compared to that in the environment just outside the crevice. In this case, metal at the site of low oxygen concentration will become anodic and corrosion will occur within the crevice.

Pitting corrosion. When protective films covering a metal, or layers of the metal, or the uniform surface of a metal break down, corrosion takes the form of pitting (Figure 2-10). The metal within the pit area is anodic and the surrounding area is cathodic.

Impingement corrosion attack. Many metals under normal conditions form protective surfaces as a result of the manufacturing process. An extremely forceful stream of water may break through this protective surface and cause corrosion to develop (Figure 2-11). Pumps are especially vulnerable to this type of attack because of cavitation—the

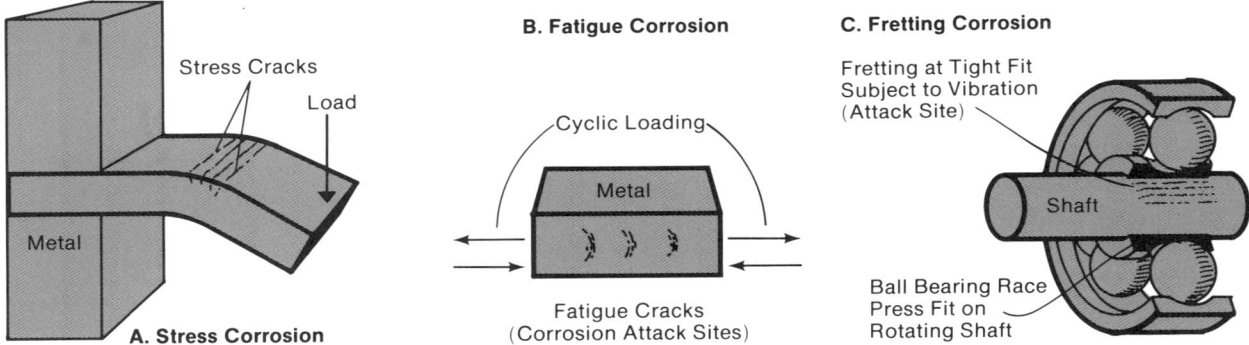

Figure 2-13   Stress, Fatigue, and Fretting Corrosion

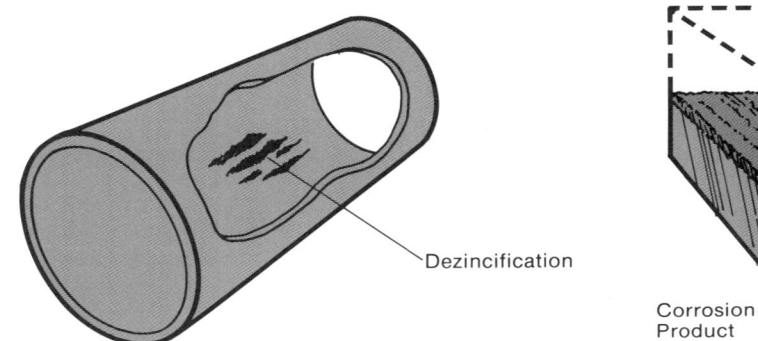

Figure 2-14   Selective Corrosion

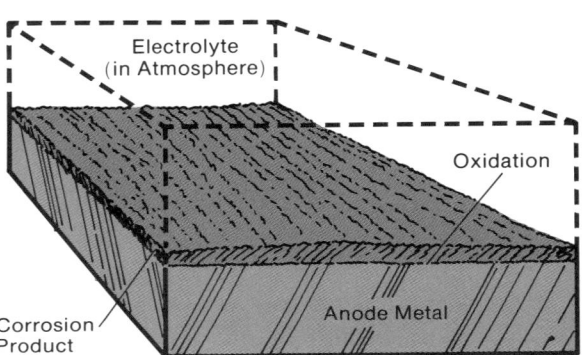

Figure 2-15   Atmospheric Corrosion

abrasive effect caused by the collapse of bubbles formed by extreme pressure differentials within the pump mechanism.

Bacteriological corrosion.   By-products of sulfate-reducing bacteria cause bacteriological corrosion (Figure 2-12). These bacteria live in environments where there is little or no oxygen and where the pH is near neutral. Their life processes give off sulfides and other sulfur compounds that are excellent electrolytes, depolarizers, and are otherwise aggressive to metal surfaces.

Soil corrosion.   Most soil corrosion occurs in soil with high electrical conductivity (low electrical resistivity), which makes the soil an effective electrolyte. Nonuniformity, chemical contamination, or differential aeration (areas of high and low oxygen content) can increase the problem. Sometimes low-resistivity soils act in combination with sulfate-reducing bacteria to create extremely aggressive conditions. Table 2-2 lists some typical soil corrosion cells that result from variances in electrolyte along a metal surface.

The pH of the soil may be significant. Acid soils with pH below 5.0 are generally aggressive. A neutral pH (6.5–7.5) does not indicate corrosive conditions unless sulfate-reducing bacteria are involved. High pH soils are not normally aggressive to ferrous metals due to their alkalinity; however, they are usually high in soil salts, which results in low soil resistivities and makes such soils good electrolytes.

Moisture is generally a requirement for soil corrosion, because dry soils make poor electrolytes. Very dry soils seldom cause corrosion problems.

Table 2-2 Typical Soil Corrosion Cells Resulting From Nonuniform Electrolyte Conditions

Anodic Area	Cathodic Area
Low oxygen	Higher oxygen
Low resistivity	Higher resistivity
Sulfate-reducing bacteria	No sulfate-reducing bacteria
Water-saturated soil	Drier soil
Organic contamination (dead vegetation, leaves, refuse)	Clean soil
Contact with highly conductive material (cinders, coal, salts)	Clean soil

Stress, fatigue, and fretting corrosion. Physical reactions and corrosion can combine to deteriorate metal and alloys, as in stress, fatigue, and fretting corrosion (Figure 2-13). Tensile stresses, cyclic stresses, or high-frequency vibrations acting independently may or may not cause material failure, but a corrosive environment will greatly enhance and accelerate the deterioration of metal in each case.

Selective corrosion. Selective corrosion singles out one element of an alloy (Figure 2-14). One material that may suffer selective corrosion is common yellow brass, which consists of approximately 30 percent zinc and 70 percent copper. In certain wet environments, zinc may be lost from the brass, leaving a porous copper mass with greatly reduced strength. This effect is termed dezincification.

Atmospheric corrosion. Atmospheric corrosion (Figure 2-15) requires atmospheric humidity and an oxidizing agent, usually oxygen. Most water treatment plants have areas where materials are constantly exposed to very humid conditions. The result is oxidation—in the case of iron or steel, rusting. In addition to oxygen, the halogens—chlorine, fluorine, iodine, and bromine—are extremely active oxidizing agents in such environments.

Atmospheric corrosion may tend to cause uniform corrosion attack. Oxidizing agents (acids, chlorine, etc.) dissolve the corrosion product layer and oxidation continues.

Glossary

(Glossary terms are defined at the back of the book.)

Anode
Bacteriological corrosion
Cathode
Cathodic protection
Concentration cell
Corrosion cell
Current
Dezincification
Electrode
Electrolyte
Electrolytic corrosion cell
Electron
Electronic path
Faraday's law
Fatigue corrosion
Fretting corrosion

Galvanic corrosion cell
Galvanic series
Impingement corrosion
Ion
Ohm's law
Passivation
Pitting corrosion
Polarization
Polar solvent
Potential (voltage)
Resistance
Return current path
Soil corrosion
Stray current
Stress corrosion
Voltage

AWWA MANUAL M27

Chapter 3

Evaluating the Potential for Corrosion

External corrosion of water utility piping and equipment is not inevitable. Certain combinations of materials in certain environments may suffer serious corrosion at a rapid rate, while other configurations may be relatively immune. For the purpose of both design and operations, it is important to be able to predict the occurrence and severity of corrosion.

For metallic corrosion to occur, four elements are required: an anode, a cathode, an electrolyte, and a conductive path for return current. Dissimilar metals and alloys are commonly found in contact in water utility installations, and even a single piece of metal can simultaneously act as anode, cathode, and return current path. In either case, the electrolyte is a key variable in determining where and to what extent corrosion will occur. For external corrosion, the electrolyte is the environment—the soil or water surrounding the equipment.

Many environmental conditions will not support corrosion. Other environments serve as a damaging electrolyte for some metals, alloys, or metallic pairs, but not for others. To evaluate an environment for a given material, experience provides the best guide. In the absence of experience, as in the case of new locations, analytical procedures should be used. If either experience or analysis shows the environment to be aggressive, corrosion prevention or control should be initiated.

This chapter details the relationships among specific materials and environmental conditions commonly found in water utility installations, then concludes with a discussion of monitoring for stray electrical currents that may cause electrolytic corrosion.

After completing this chapter, the reader should be able to
- understand the influence of environment on the corrosion process;
- realize that a given environment may cause corrosion in one material but not in another;
- be alert to many of the environmental conditions that cause corrosion;
- have a general understanding of the systems used to determine whether soils are corrosive to various materials;
- understand how stray direct current can affect underground metallic structures, and recognize some of the common sources of stray current.

EFFECTS OF THE CHEMICAL ENVIRONMENT ON COMMON WATER PIPE MATERIALS

The effects of soil, water, and air on a pipe vary greatly depending on the pipe material. This section discusses environmental effects and evaluation procedures for materials commonly used in water utility pipe and appurtenances.

Gray and Ductile Cast-Iron Pipe

Several soil evaluation procedures for gray and ductile cast-iron pipe installations have been developed and used to varying extents. Because of the complexity of underground corrosion, no system is universally applicable, and no system can replace the need for expertise in the field of corrosion control. The evaluation system originally developed and recommended by the Cast Iron Pipe Research Association,* commonly referred to as the 10-point system, is discussed in detail in the following text. This soil evaluation procedure is specific to gray and ductile cast-iron pipe and should not be applied to other materials.

Appendix A of ANSI/AWWA C105/A21.5-82, Standard for Polyethylene Encasement for Ductile-Iron Piping for Water and Other Liquids, covers the soil-survey tests, observations, and interpretations that comprise the 10-point system. The soil evaluation system was made public in 1968 and adopted as an appendix to the standard in 1972. It was based on field experience with operating cast-iron pipelines, where soil-test results were recorded along with the age of the pipe and its condition, including extent and type of corrosion. Most of the experience was with gray cast-iron pipe. When ductile-iron pipe became a production item, it was necessary to determine whether the same tests and evaluation could also be accurately applied to it. Field research has shown that the overall corrosion rates of ductile and gray cast iron are such that the soil evaluation system can be applied to both with equal accuracy.

The evaluation procedure is based on information from five tests and observations: soil resistivity, pH, redox potential, sulfides, and moisture. For a given soil sample, each result is evaluated and assigned points according to its contribution to corrosivity. The points for all five areas are totaled. If the sum is 10 or more, the soil is corrosive to gray or ductile cast-iron pipe, and corrosion will likely occur unless protective measures are taken (as discussed in Chapter 4). Table 3-1 lists the points assigned to results from the various tests. The following paragraphs give additional detail on each area of evaluation.

Soil resistivity. The soil-resistivity test yields the reciprocal of conductivity; a low resistivity of soil means that it will serve well as an electrolyte. Resistivity, reported in ohm-centimetres (Ω-cm), represents the average of the electrical resistances across each cubic centimetre of soil in a given volume. The test may be accomplished in several ways. For an average resistivity from near the ground surface to pipe depth or below, the 4-pin system is useful (Figure 3-1).

The 4-pin system has limitations because it may average dry top soil with wetter subsoil, contaminated soil with clean soil, etc. Use of a single soil probe (Figure 3-2) is preferred by some corrosion technicians for field tests. The probe is suitable for specific readings at various soil depths, enabling the surveyor to search out the lowest-resistivity soil that may come in contact with pipe after it has been buried.

Because soil moisture affects resistivity, and moisture in the field will vary unpredictably, it is advisable to remove a soil specimen from pipe depth for laboratory testing. The quad-box (Figure 3-3) or a similar unit makes it possible to water-saturate the soil, putting available salts in solution and simulating the worst condition likely to occur after pipe installation.

---

*The Cast Iron Pipe Research Association (CIPRA) became the Ductile Iron Pipe Research Association (DIPRA) in 1979.

22 EXTERNAL CORROSION

Table 3-1 Soil-Test Evaluation for Gray or Ductile Cast-Iron Pipe (10-Point System)*

Soil Characteristics	Points
Resistivity—$\Omega\text{-}cm$†	
<700	10
700–1000	8
1000–1200	5
1200–1500	2
1500–2000	1
>2000	0
pH	
0–2	5
2–4	3
4–6.5	0
6.5–7.5	0‡
7.5–8.5	0
>8.5	3
Redox potential	
>+100 mV	0
+50 to +100 mV	3.5
0 to +50 mV	4
Negative	5
Sulfides	
Positive	3.5
Trace	2
Negative	0
Moisture	
Poor drainage, continuously wet	2
Fair drainage, generally moist	1
Good drainage, generally dry	0

*Ten points—corrosive to gray or ductile cast-iron pipe, protection is indicated.
†Based on single probe at pipe depth or water-saturated soil box.
‡If sulfides are present and low or negative redox-potential results are obtained, three points shall be given for this range.

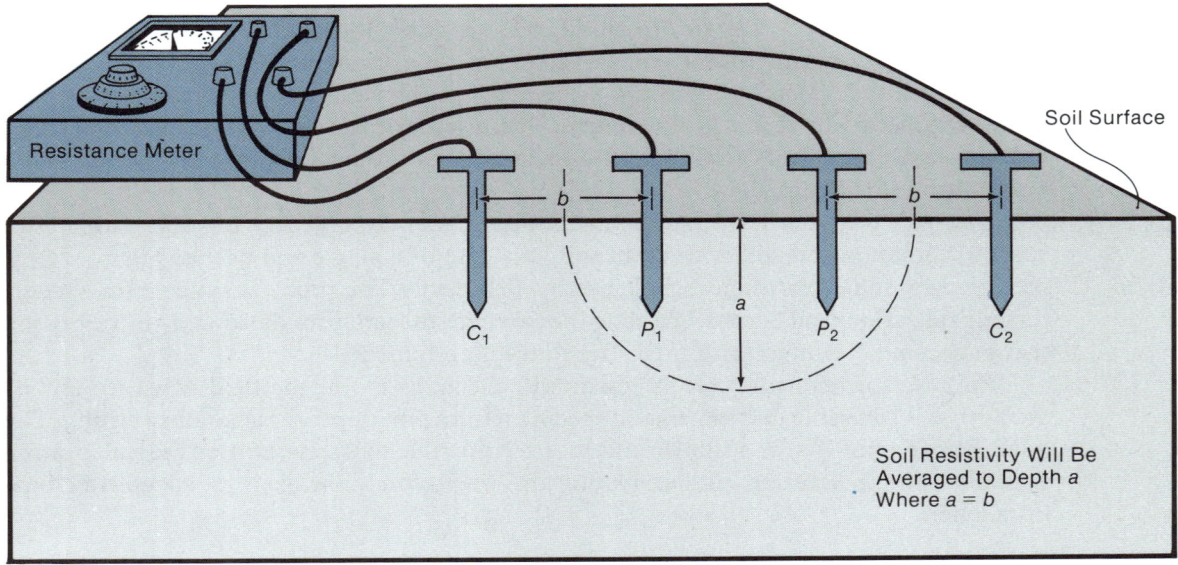

Figure 3-1 The 4-Pin System of Soil-Resistivity Testing

POTENTIAL FOR CORROSION 23

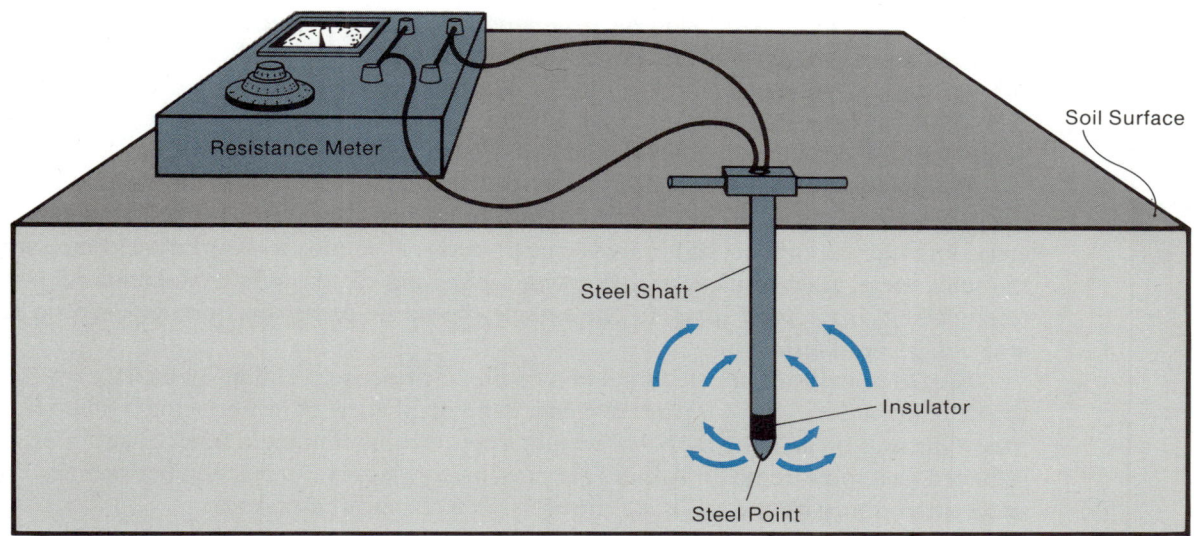

Figure 3-2 Use of a Single Probe for Testing Soil Resistivity

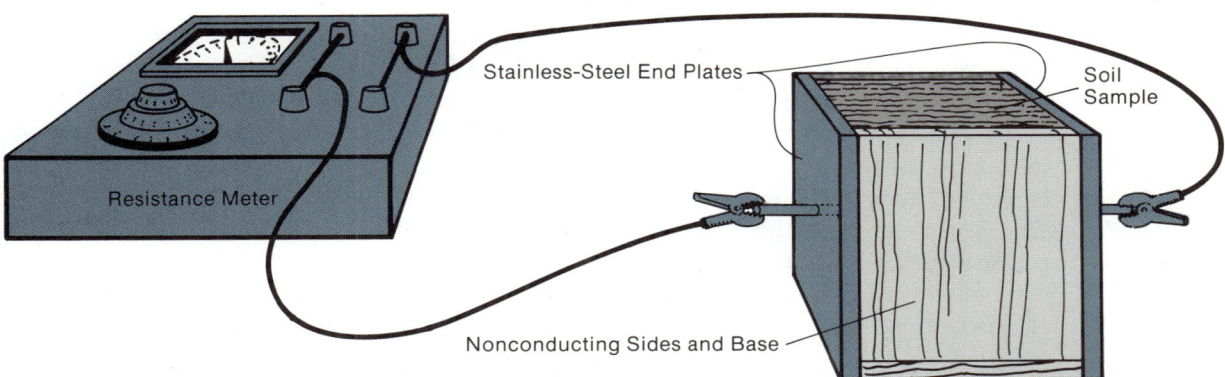

Figure 3-3 Quad-Box for Testing Soil Resistivity of a Water-Saturated Soil Sample

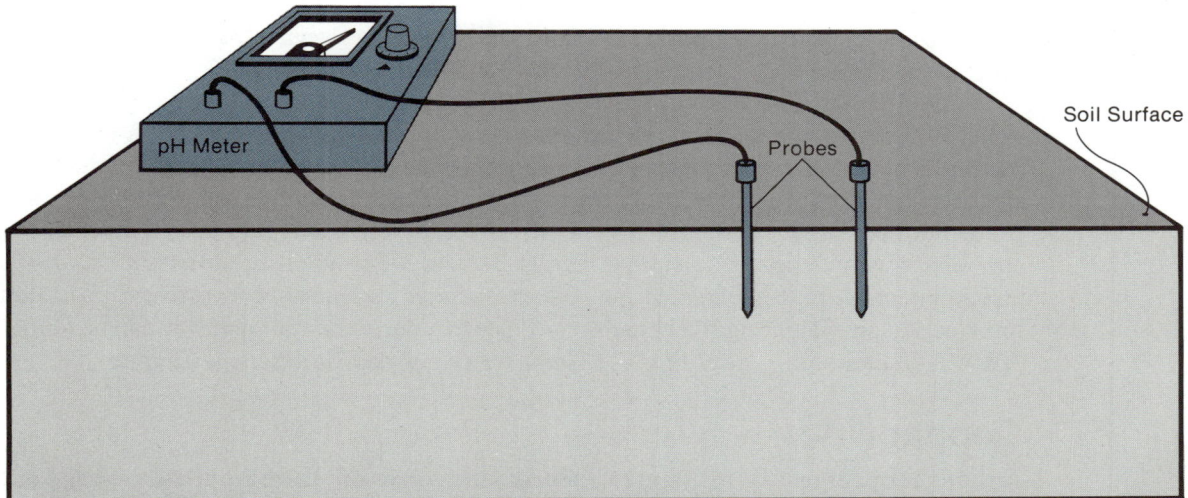

Figure 3-4 Testing Soil pH

pH. Soils with a pH below 4.0 generally serve well as an electrolyte, are high in total acids, and have a record of being aggressive. Neutral pH (6.5–7.5) indicates that the soil will support sulfate-reducing bacteria if other characteristics are suitable. Soils with a high pH (8.5–14.0) are generally high in dissolved salts and usually exhibit a low resistivity. The equipment for measuring pH is illustrated in Figure 3-4.

Oxidation-reduction (redox) potential. A test of the oxidation-reduction potential indicates the degree of aeration of the soil. Low or negative results show that the soil is anaerobic and will support sulfate-reducing bacteria. The redox test can be performed with the same meter used to measure pH, using a platinum electrode in conjunction with the reference electrode used for pH. The sample should be protected from exposure to the atmosphere until tested.

Sulfides. If sulfides are found in the soil, the presence of sulfate-reducing bacteria is likely. The sulfides test is qualitative and is accomplished by introducing a solution of 3-percent sodium azide in $0.1N$ iodine into a test tube containing a small quantity of soil removed from pipe depth. If sulfides are present, they catalyze a reaction between sodium azide and iodine with the release of nitrogen. The chemical reaction is

$$\underset{\text{Sodium Azide}}{2NaN_3} + \underset{\text{Iodine}}{I_2} \rightarrow \underset{\text{Sodium Iodide}}{2NaI} + \underset{\text{Nitrogen}}{3N_2}$$

Moisture. The sample should be protected from exposure to the atmosphere until tested. Since the soil moisture may vary throughout the year, general drainage characteristics are recorded rather than specific moisture content.

Experience. In addition to the analytical tests just described, which will give clear indications of the ability of the soil environment to cause gray and ductile cast-iron pipe corrosion, notes on prior experience in the area are extremely important. In many cases, experience can yield the best predictions of soil corrosivity. As general information, a national survey has indicated that about 5 percent of the land area where water distribution systems are installed in the United States contains soil that is corrosive to gray and ductile cast-iron pipe.

Steel Pipe

In the discussion that follows, it should be borne in mind that "steel pipe" indicates a product with a composite wall system. Bare steel pipe is not recommended and is not normally used for buried installation in water systems. In current practice, the steel pipe wall system consists of the structural steel core sandwiched between a dielectric coating and a protective lining.

The potential for corrosion of the exterior of steel pipe is difficult to judge because of the variety of environments encountered. Assessing the corrosion hazard is an engineering matter. Soil chemical and physical analyses, pH, moisture content, and existence of stray electrical currents are important factors that can aid in determining the need for corrosion protection. Resistivity of the soil is generally the most important parameter for judging soil corrosivity with respect to steel pipe. Tables 3-2 and 3-3 summarize the extent of corrosion attack on steel pipe as related to soil characteristics and soil resistivity. Additional information on evaluating steel pipe's resistance to environmental corrosion can be found in AWWA Manual M11, *Steel Pipe—A Guide for Design and Installation*, Chapter 10.

Copper Pipe

Although copper may be resistant to underground corrosion, there are certain soil and other environmental conditions that can cause it to deteriorate. Specific soil chemistries that may cause problems include a high content of organic matter of high alkalinity, in which the ratio

Table 3-2 Soils Grouped in Order of Corrosive Action on Steel

Group I—Lightly Corrosive
Aeration and drainage good. Characterized by uniform color and no mottling anywhere in soil profile and by very low water table. Includes:
1. Sands or sandy loams
2. Light, textured silt loams
3. Porous loams or clay loams thoroughly oxidized to great depths

Group II—Moderately Corrosive
Aeration and drainage fair. Characterized by slight mottling (yellowish brown and yellowish gray) in lower part of profile (depth 18–24 in.) and by low water table. Soils would be considered well drained in an agricultural sense, as no artificial drainage is necessary for crop raising. Includes:
1. Sandy loams
2. Silt loams
3. Clay loams

Group III—Badly Corrosive
Aeration and drainage poor. Characterized by heavy texture and moderate mottling close to surface (depth 6–8 in.) and with water table 2–3 ft below surface. Soils usually occupy flat areas and would require artificial drainage for crop raising. Includes:
1. Clay loams
2. Clays

Group IV—Unusually Corrosive
Aeration and drainage very poor. Characterized by bluish-gray mottling at depths of 6–8 in. with water table at surface, or by extreme impermeability because of colloidal material contained. Includes:
1. Muck
2. Peat
3. Tidal marsh
4. Clays and organic soils
5. Adobe clay

Table 3-3 Relationship of Soil Corrosion to Soil Resistivity

Soil Class	Description	Resistivity Ω-cm
1	excellent	10 000–6000
2	good	6000–4500
3	fair	4500–2000
4	bad	2000–0

of chloride and carbonate to sulfate is high; high concentrations of organic and inorganic acids; poor aeration, which supports sulfate-reducing bacteria activity; and a high chloride, sulfide, and/or ammonia content. Corrosion of copper by a soil can be aggravated by the application of fertilizer and heavy lawn sprinkling, especially if the soil is poorly drained. Copper should not be embedded directly in cinders or in tidal marshes where it may be subjected to attack by sulfur compounds.

Copper in contact with concrete is often cathodic to nearby copper exposed to soil, causing the pipe in the soil to corrode. Premature failure of the copper at the concrete/soil interface in homes constructed on concrete slabs has occasionally been reported.

A hazard to underground copper is the practice of using metal water service lines as a ground for alternating-current electrical systems. If copper service lines connected to nonconducting (for example, asbestos–cement) mains are used as a ground, corrosion damage can sometimes occur where current leaves the copper, even if the current is alternating. The alternating current evidently is rectified (changed to direct current, which will cause corrosion) by copper oxide films on the pipe surface, under certain soil, pH, and electrical field intensity conditions.

Although seldom a problem for a water utility, it is interesting to note that underground domestic hot-water copper pipes can be anodic to the copper cold-water pipes. If the two systems are electrically connected, as can occur at the hot-water heater, the hot-water pipes can experience accelerated corrosion by the mechanism of thermogalvanic action. This can be overcome by insulating the hot- and cold-water lines from each other with an isolating union.

Concrete Cylinder Pipe

Concrete cylinder pipe is a general designation given to pipe manufactured with a watertight steel cylinder and reinforcing or prestressing wire, all embedded in a rich concrete or mortar encasement. There are presently three types being produced in the United States and Canada: reinforced concrete cylinder pipe (Figure 3-5), prestressed concrete cylinder pipe (Figures 3-6 and 3-7), and pretensioned concrete cylinder pipe (Figure 3-8). These types are described in detail in AWWA C300, Standard for Reinforced Concrete Pressure Pipe, Steel Cylinder Type, for Water and Other Liquids; AWWA C301, Standard for Prestressed Concrete Pressure Pipe, Steel-Cylinder Type, for Water and Other Liquids; and AWWA C303, Standard for Reinforced Concrete Pressure Pipe—Steel Cylinder Type, Pretensioned, for Water and Other Liquids.

Concrete cylinder pipe has demonstrated consistent long life when properly installed in most natural environmental conditions. This is attributable to the protection Portland cement affords embedded steel. The hydrated cement is chemically basic, having a pH of about 12.5. At this pH, an oxide film forms on the embedded steel surfaces, which passivates the steel. The passivated steel will not corrode unless a corrosion cell condition is created having a voltage differential of more than 1.5 V. This voltage is higher than that found in concentration or aeration cells formed by natural installation environments.

If a galvanic cell is created with concrete-coated, uncorroded steel as one electrode and copper–copper sulfate as the other, the voltage differential created will be in the range of 0 to −300 mV. This corrosive driving voltage is 300 to 500 mV less than would be experienced with bare or organically coated steel. If a similar cell is created, but with concrete-coated steel that has been induced to corrode, the measured voltage differential will be considerably different. The difference between measurable voltages created by corroding and noncorroding steel can be used to monitor for possible corrosion activity beneath the concrete coating.

In the majority of environments, no protective requirements are necessary for concrete cylinder pipe. However, there are unusual circumstances where precautionary measures should be taken to ensure pipeline integrity. These circumstances include some chloride environments, which can depassivate steel; cathodic interferences from man-made structures; severely acidic soils, which can chemically attack the coating; atmospheric exposure, which can deteriorate the coating; and soils high in sulfate concentration. Each of these conditions is discussed in the following paragraphs.

Chlorides. The breakdown of passivation and possible steel corrosion can occur in sound concrete only if certain negative ions (anions) are present at the steel surface. The only anion that causes practical concern is the chloride ion. When exposed to a high concentration of chloride ions and oxygen, the iron in steel will corrode. The threshold chloride ion concentration required at the steel surface to initiate corrosion is quite high—about 0.02 mol/L (approximately 700 mg/L). However, in the absence of impressed (externally generated) voltages, oxygen must be present at the steel surface to sustain corrosion. Even in high-chloride environments such as continuously submerged ocean outfalls (approximately 20 000 mg/L chloride), concrete cylinder pipe has shown no evidence of corrosion of embedded steel. This is due to the extremely low rate of oxygen diffusion through the saturated mortar coatings, even though the dissolved-oxygen content of the seawater may be relatively high.

POTENTIAL FOR CORROSION   27

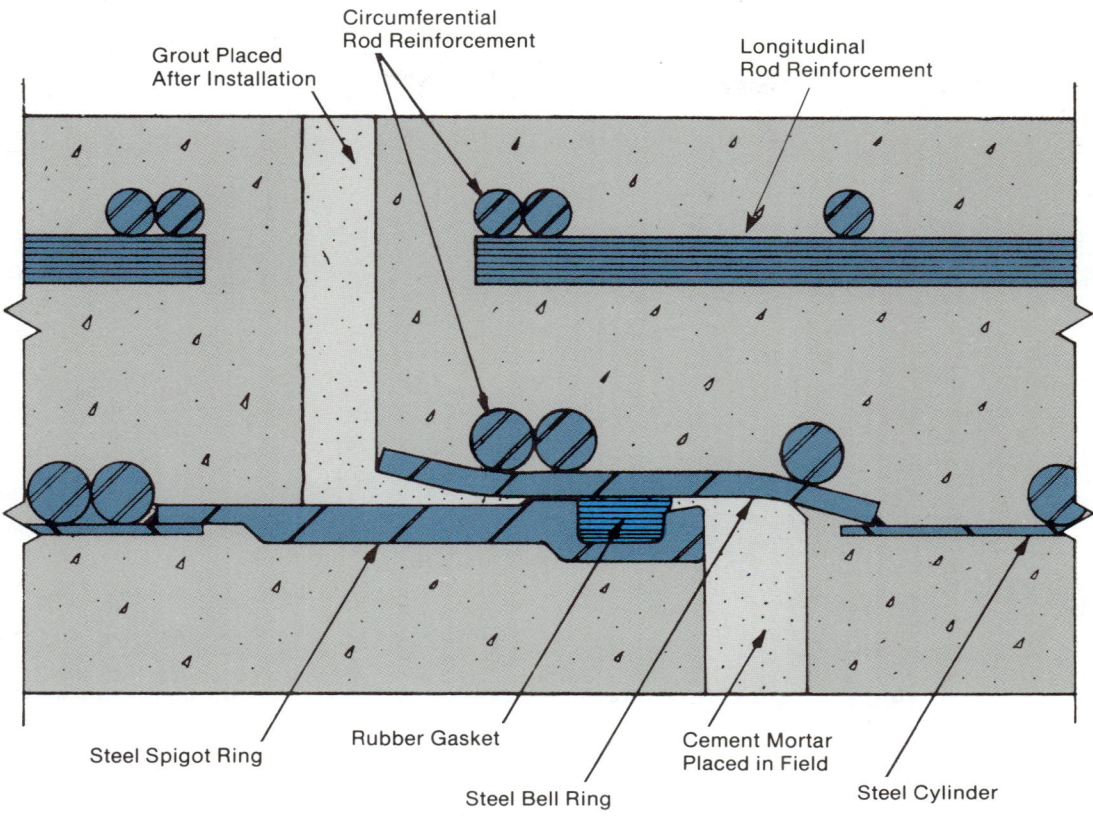

Figure 3-5   Reinforced Concrete Cylinder Pipe With Rubber and Steel Joint (AWWA C300)

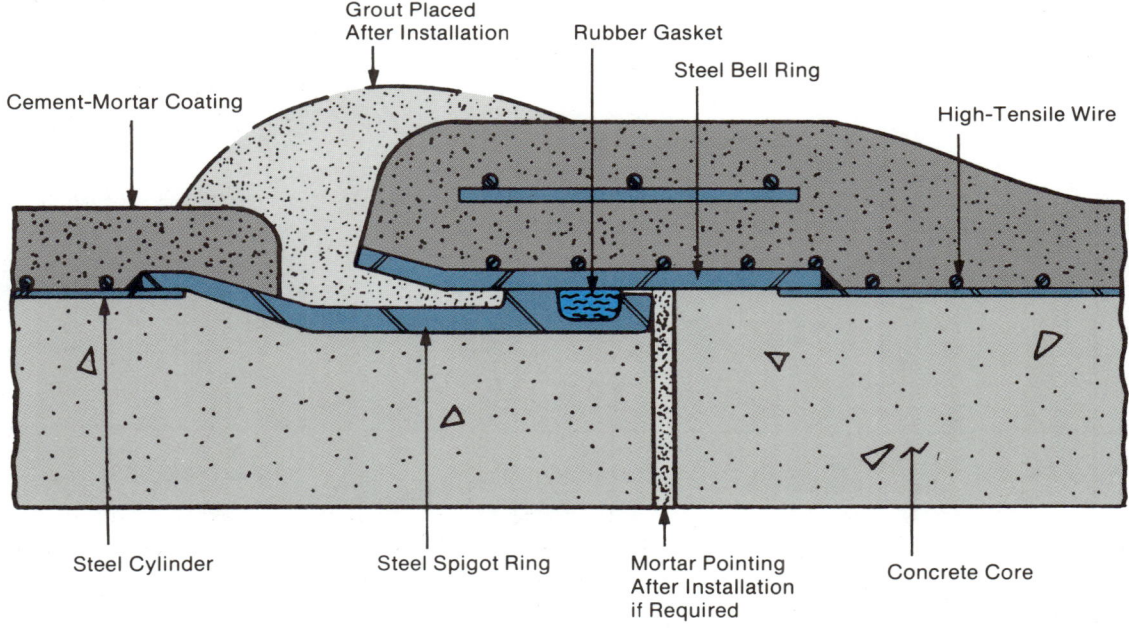

Figure 3-6   Prestressed Concrete Cylinder Pipe With Rubber and Steel Joint (AWWA C301)

28　EXTERNAL CORROSION

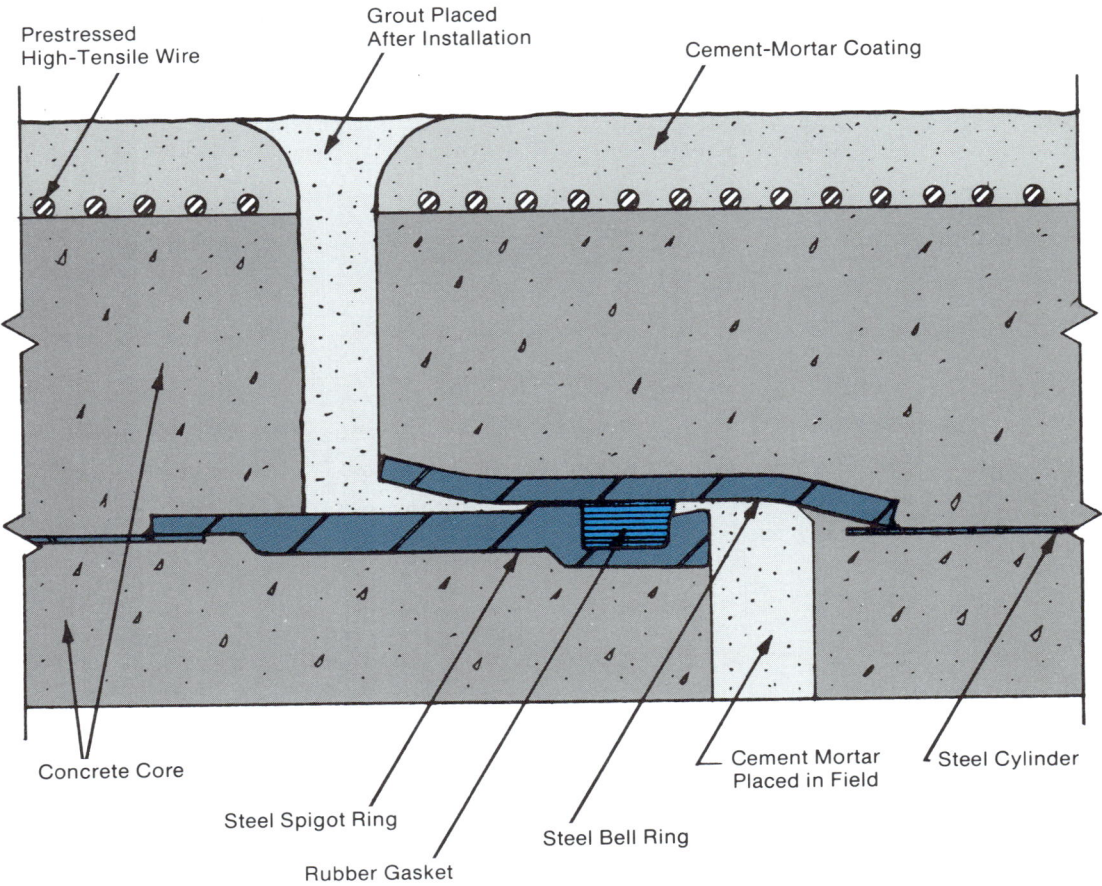

Figure 3-7　Prestressed Concrete Embedded Cylinder Pipe With Rubber and Steel Joint (AWWA C301)

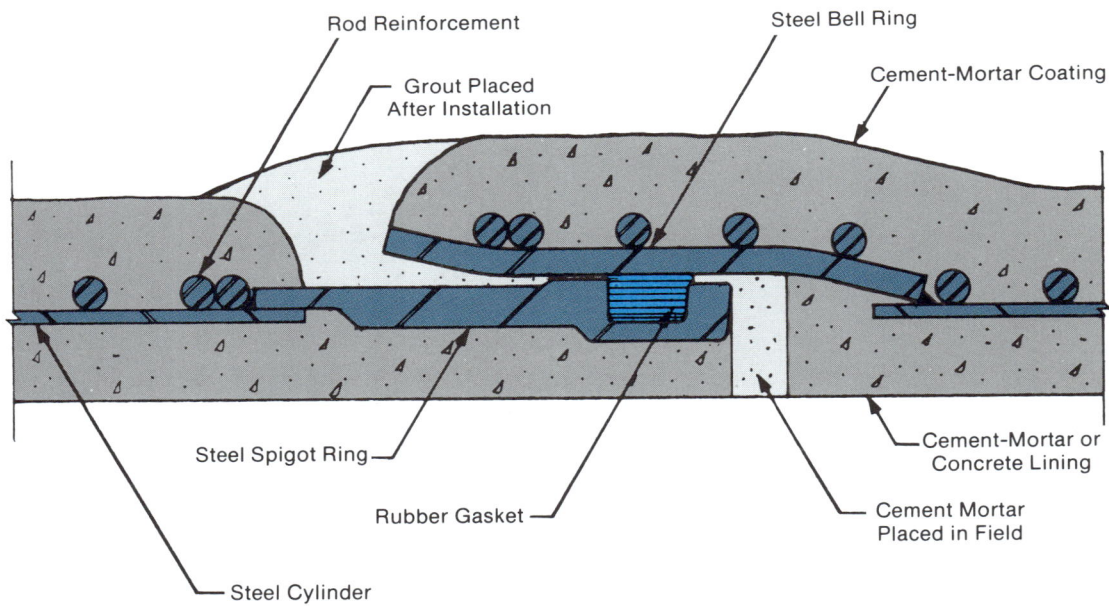

Figure 3-8　Pretensioned Concrete Cylinder Pipe With Rubber and Steel Joint (AWWA C303)

In high-chloride environments where the mortar coating cannot be assumed to remain saturated, specific consideration should be given to both chloride-ion content of the soil and to soil resistivity. If chloride ion concentration in the soil exceeds 1000 ppm and soil resistivity, as measured in the maximum natural moisture state, is lower than 1000 Ω-cm, it is wise to protect the exterior surface with a coal-tar coating or to bond all pipe joints and provide a system for monitoring corrosion activity.

Cathodic interference. The section on stray current at the end of this chapter covers evaluation of the environment for corrosive conditions that could result from the cathodic protection of other systems in the area of the pipeline.

Acidic soils. Significant chemical attack of concrete cylinder pipe in acid soils is extremely rare, usually occurring only in soils affected by contamination such as cinders, mine wastes, or industrial dumps. Pipelines installed in low-pH, high-total-acidity soils in the presence of a fluctuating water table may suffer corrosion. Generally speaking, no problems will occur in soil having a pH of 5 or higher where there is a low probability of groundwater movement with respect to the pipe. Where pH values below 5 are encountered and where considerable movement of groundwater with respect to the pipe is anticipated, corrosion may occur in the absence of protective measures, such as a special backfill or exterior coatings (discussed in Chapter 4).

Atmospheric conditions. Although concrete cylinder pipe is normally buried below the ground, it occasionally becomes necessary to make an aboveground installation. The atmospheric environment is drastically different from buried conditions. Atmospheric conditions bring alternate wetting and drying and rapidly changing temperatures between night and day. No longer is the pipe surrounded by moist or slowly changing temperatures within a narrow range of approximately 30°F (17°C)—the pipe surface temperatures may fluctuate from a winter low of below freezing to a hot summer day of well over 100°F (38°C). The temperature fluctuations and the expansion of freezing water in concrete produce stresses and strains in the pipe wall that may deteriorate the cement-mortar coating. In general, the conditions associated with atmospheric exposure can be expected to reduce pipe life unless additional protective measures are taken.

Sulfate soils. Soils containing high concentrations of sodium, magnesium, and calcium sulfates are designated sulfate soils (often incorrectly termed alkali soils) and are chiefly found in parts of western Canada and the western and southwestern United States. Under certain circumstances, these soils can be aggressive to conventional, field-placed concrete structures, particularly those in contact with the soil but with partial atmospheric exposure. Although concrete cylinder pipe can also be attacked by sulfate soils, the pipe has generally exhibited excellent sulfate resistance. This is attributed primarily to the high cement content of the mortar coating, averaging in excess of 10 sacks (940 lb [430 kg]) per cubic yard.

Asbestos–Cement Pipe

Asbestos–cement (AC) pipe is available in two basic types, which are distinguished by the free lime content of the pipe. Type I pipe has no limit on uncombined calcium hydroxide; type II pipe has 1 percent or less uncombined calcium hydroxide.* In the United States, only type II AC pipe has been produced for the past several decades. Type II pipe contains, in addition to asbestos fiber and Portland cement, approximately 30 percent finely ground silica. The curing process is completed in high-pressure steam autoclaves, where the temperature is approximately 300°F (150°C). The series of compounds known as hydrogarnets, which result from steam curing of cement products, are very stable and highly resistant to the action of sulfate solutions.

---

*Tests for uncombined calcium hydroxide are made in accordance with ASTM C500, available from the American Concrete Institute, PO Box 19150, Redford Station, Detroit, MI 48219.

Guidelines for the use of AC pipe under various soil conditions are established in AWWA C400, Standard for Asbestos-Cement Distribution Pipe, 4 in. Through 16 in. (100 mm Through 400 mm) NPS, for Water and Other Liquids; AWWA C402, Standard for Asbestos-Cement Transmission Pipe, 18 In. Through 42 In. (450 mm Through 1050 mm), for Potable Water and Other Liquids; and ASTM C500, Standard Methods for Testing Asbestos-Cement Pipe. These guidelines establish parameters that are intended to define conditions where indefinite life expectancy can be anticipated.

Table 3-4 shows guidelines for the use of AC pipe in acidic soils based on minimum pH factors alone. Asbestos-cement pipe may or may not perform satisfactorily in acid soil environments having pH values below those listed in this table. To determine the suitability of AC pipe in soils having lower pH values, each situation should be evaluated individually, taking into consideration all aspects of the soil environment that affect corrosiveness to AC pipe.

Table 3-5 classifies aggressiveness to AC pipe of soluble sulfates in water and soils. Type I AC pipe will be attacked to various degrees by all but the nonaggressive levels of sulfate concentrations in waters and soils. Type II AC pipe is resistant to all levels of soluble sulfates.

Table 3-4 Guidelines for Use of Asbestos-Cement Pipe Based on pH of Acidic Soils

Water Conditions Within Soil Environment	Minimum pH of Acidic Soils When Using Asbestos-Cement Pipe	
	Type I	Type II
Essentially quiescent	5.0	4.0
Mildly fluctuating	5.5	5.0
Rapidly moving or grossly cyclic	6.3	5.5

Table 3-5 Corrosion Guidelines for Asbestos-Cement Pipe for Soluble Sulfate in Water and Soils

Sulfate Aggressiveness	SO_4 in Soil *ppm*
Nonaggressive	1000 and less
Mildly aggressive	1000 to 2000
Moderately aggressive	2000 to 20 000
Highly aggressive	20 000 and above

Thermoplastic Pipe

Polyvinylchloride (PVC), polyethylene (PE), and polybutylene (PB) are the thermoplastic materials most commonly used in water distribution. These materials do not corrode in the sense that metals do. Being nonconductors, they are immune to corrosion by galvanic or electrochemical effects. They are unaffected by polar active materials, such as aqueous acids, bases, and salts. Consequently, linings, coatings, and cathodic protection are not required with thermoplastic piping. However, certain substances that are seldom encountered in municipal water distribution environments can adversely affect thermoplastic pipe's structural properties through direct chemical attack, solvation, or environmental stress cracking. Pipeline contact with such substances can occur in instances of subsurface contamination—for example, leaking gasoline storage tanks.

Direct chemical attack results in a progressive change or breakdown of the plastic's molecular structure. It is brought about only by very strong oxidizers, acids, bases, or long-term exposure to ultraviolet light. Solvation is the absorption of an organic solvent.

Depending on the plastic, the solvent's nature, and the quantity of the solvent absorbed, the effects of solvation can range from a slight swelling of the plastic with little loss of properties to a complete solution. Environmental stress cracking is the development and growth of cracks when the plastic is simultaneously under stress and exposed to certain organic liquids. PVC is not stress-crack sensitive, but certain grades of PE can be.

If thermoplastic pipe is considered for use where exposure to chemical reagents, solvents, and other organic substances may occur, the pipe's suitability for the specific exposure should be determined. For example, the use of PVC pipe is not advised when it will be exposed to strong PVC solvents, such as methyl ethyl ketone, acetone, or tetrahydrofuran. For a complete listing of PVC pressure pipe's environmental resistance, consult AWWA Manual M23, *PVC Pipe—Design and Installation*, Sec. 1-2, Corrosion and Chemical Resistance. Information on the environmental resistance of PE and PB pipe can be obtained from the Plastics Pipe Institute* or from polyethylene pipe manufacturers.

STRAY CURRENTS

Direct-current electricity flowing through the soil in the area of a metal pipeline or structure can induce electrolytic corrosion of the metal or alloy. The electricity is referred to as stray current because it has strayed away from the path intended for it in the circuit where it originated. Stray current is primarily a problem for long, electrically continuous pipelines, which can form a lower resistance pathway for the current than is found in its original circuit. Electrically segmented pipelines (those using rubber-gasket joints) are usually not vulnerable to stray-current corrosion.

When planning a new pipeline, or if stray direct current is suspected of causing corrosion, the first step in analyzing the problem is to review the possible sources of direct current. Some common sources include impressed-current cathodic-protection systems, direct-current powered trains or streetcars, arc-welding equipment, and direct-current transmission systems. If a local corrosion-control coordinating committee exists, its members are usually well informed on such sources. If there is no such committee, other utility operators should be consulted.

Three electrical tests are commonly used for the analysis of stray current in the field. Each test is performed by measuring a voltage differential. The pipe-to-soil (structure-to-environment) test measures the voltage between the pipe and the soil. The current-span (IR-drop) test measures the voltage between two points on the pipeline. The ground-voltage gradient test measures the voltage between two points in the soil. When test leads are placed at appropriate intervals along the pipeline, these tests allow the line to be monitored both for the existence of environmental stray currents and, to some extent, for the existence of active corrosion. The tests and observations are best applied to long, electrically continuous metal pipelines. Bonding of rubber-gasketed joints may be necessary if the line is to be monitored for potential stray current or cathodically protected.

The following sections briefly describe each test and its interpretation. Note that any attempt to monitor a pipeline requires the maintenance of careful records showing the location of outlets, connections to other lines, pipeline appurtenance structures (such as air valves and manholes), and the proximity of foreign lines or structures that could influence monitoring data.

Pipe-to-Soil Potential

Pipe-to-soil potential (also called structure-to-environment potential) is the potential between the metal and a reference electrode placed in contact with the earth. The reference

---

*Plastics Pipe Institute, 355 Lexington Ave., New York, NY 10017.

32 EXTERNAL CORROSION

cell is half of a battery. In the commonly used copper–copper sulfate reference electrode, the "half-cell" consists of a copper rod in a saturated solution of copper sulfate crystals. The other half of the cell is the metal in the earth electrolyte.

The reference cell has a stable, constant potential against which the voltage of another metal can be measured, similar to measuring elevation with reference to sea level. When the potential between a pipe and reference electrode in contact with the earth is measured, the potential includes not only the voltage between the pipe and reference cell but also IR drop voltages created by currents flowing through the resistance of the earth. This measurement is the most useful test in evaluating stray-current effects and cathodic protection effectiveness.

When a pipe is connected to the positive terminal of a digital voltmeter and the reference cell is connected to the negative terminal, the meter will give a negative reading, indicating that the pipe is negative to the reference electrode. A negative or minus sign should be placed in front of the voltage reading. Note, however, that an analog type voltmeter or potentiometer would read backwards with these connections; the reference cell must be hooked to the positive terminal and the pipe to the negative terminal so that the needle will read upscale. Under these circumstances, a negative sign must be added to show the pipe is negative to the reference cell even though the analog meter has moved in a positive direction. Figure 3-6 shows the arrangement for measuring pipe-to-soil potential with a digital voltmeter.

The best type of connection is an insulated test lead, brazed or thermite-welded to the pipe. However, connection may be made to a valve, blowoff, or other accessible appurtenance if it is certain that it is electrically continuous with the pipe.

Pipe-to-soil potential tests performed with the reference electrode placed at intervals along the centerline of an electrically continuous pipeline may give varying readings, indicating anodic (corroding) and cathodic areas of the pipe. Readings taken with the

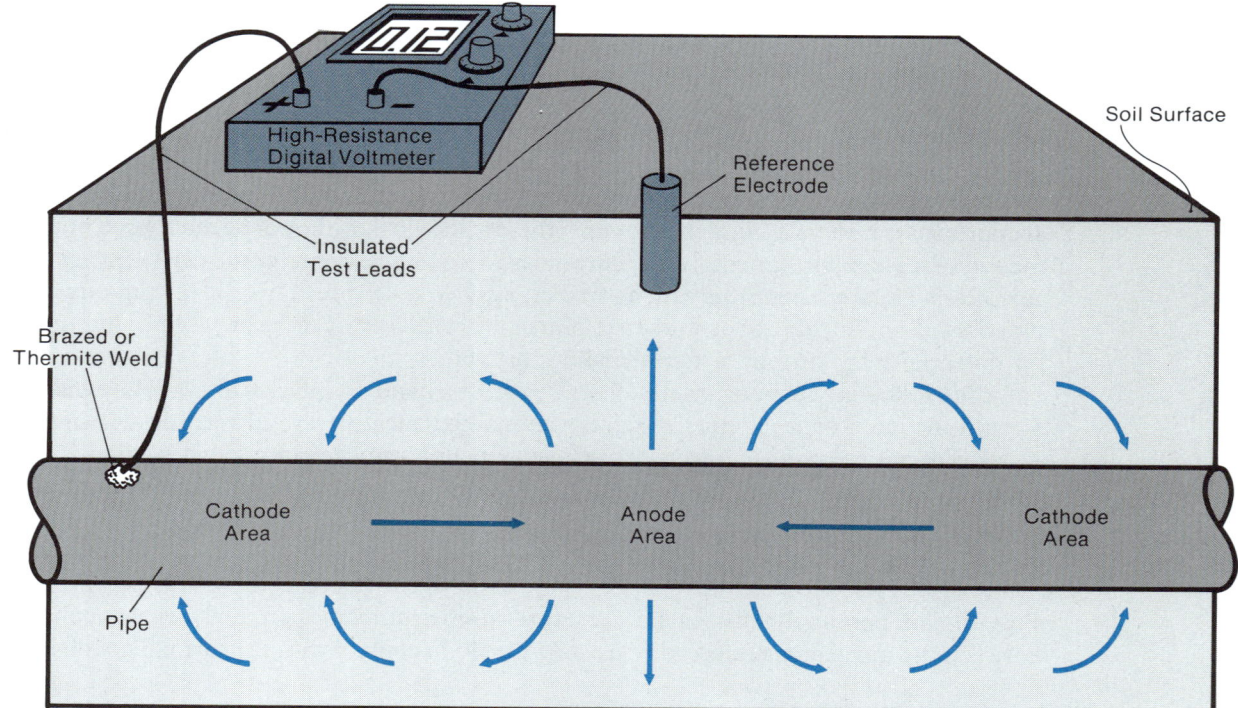

Figure 3-9 Testing for Pipe-to-Soil Potential

reference electrode offset from the centerline of the pipe can also help to locate anodic areas. Offset readings identical to the pipe alignment readings indicate neither cathodic nor anodic conditions, and if consistent, rule out the existence of stray current.

Pipe-to-soil potentials become less negative as a reference cell is moved closer to the pipe, then current is flowing toward the pipe and the pipe is cathodic. If pipe-to-soil potentials become less negative as the reference cell is moved away from the pipe, then the pipe is anodic and tends to corrode. This is true for natural galvanic corrosion cells such as between dissimilar metals and alloys or anodic and cathodic areas on the same metal or alloy and where the pipe is under cathodic protection. But, if the pipe is losing current due to stray current, the pipe-to-soil potential close to the pipe will be less negative (may even go positive) than potentials farther from the pipe.

Current Span

Current-span (IR-drop) measurements are useful for identifying stray or long-line currents on a continuous pipeline and for locating suspected corrosion areas. Current span is determined by measuring the voltage between two points on a pipeline, then calculating how much current is required to induce that potential in a line of the given size and material. The test procedure uses a high-resistance voltmeter (with millivolt ranges), with the terminals connected to test leads from the pipe. For most steel pipe with welded joints, current flow in amperes can be determined by the formula

$$I = KV \qquad (3\text{-}1)$$

Where:
    I = current flow between the two points (A)
    V = measured voltage differential (mV)
    K = 4.0 W/L, a conversion factor indicating the resistance of the pipe (A/mV)
        W = the weight-per-length of the steel pipe (lb/ft)
        L = length of the span (ft)
        (calculation of K assumes welded pipe joints, steel resistivity of 15.5 $\mu\Omega$-cm, and steel density of 490 lb/ft^3).

Similar calculations can be made for other types of metallic pipelines. Current in a span can also be determined by applying a direct current opposite to the stray current between the two test points, then adjusting the applied current until the measured potential between the test points is zero. The applied countercurrent is equal to the current flowing along the pipe. A series of current-span tests along a pipeline will help locate areas of discharge or assimilation of stray current.

Ground-Voltage Gradients

Testing ground-voltage gradients helps to determine whether there is current flowing through the soil in the vicinity of an existing or potential underground structure. It also can indicate the direction of any current flow and thus help locate its source. When plotted along the pipe alignment, results of the tests may even be helpful in locating suspected anodic areas on a pipeline.

Ground-voltage gradients are tested with a high-resistance millivolt meter and two matched copper–copper sulfate half-cells. Half-cells may be checked by placing them side by side into the soil surface. A zero millivolt reading shows that they are properly matched.

With the half-cells placed into the soil surface at a selected spacing, any millivolt potential is recorded. If such a potential exists, presence of earth current is implied. The gradient is accomplished by making a series of such readings in a straight line. "Leap frogging"—leaving one half-cell in place and carrying the other past it to a new location of equal spacing along the alignment of the study—makes the test procedure go more quickly, but care must be taken to reverse the sign for each new reading. Significant variations in readings may indicate anodic and cathodic sections of the pipeline if carried out over the pipe alignment. The test can also confirm the existence of such areas that were located using pipe-to-soil tests.

Glossary

(Glossary terms are defined at the back of this book.)

Conductivity
Copper–copper sulfate electrode
 (reference electrode, or CSE)
Current span
Ground-voltage gradients
Oxidation-reduction potential
 (redox potential)
pH
pH meter
Pipe-to-soil potential
Resistivity
Resistivity meter
Single probe
Soil box
Soil resistivity
Sulfides

AWWA MANUAL M27

Chapter **4**

Preventing or Reducing Corrosion

Where tests and observations indicate that corrosion will be a problem in a new or existing installation, steps must be taken to reduce the severity of the corrosion or eliminate it entirely. A variety of methods and combinations of methods are available for corrosion control. All are based on the principle, discussed in detail in Chapter 2, that four elements are required to support corrosion—an anode, a cathode, an electrolyte, and a return current path. If any one of these elements can be eliminated, deactivated, or isolated from the others, then corrosion will be prevented.

This chapter discusses several corrosion-control systems used by water utilities, including coatings, cathodic protection, polyethylene encasement of ductile iron, selection of corrosion-resistant materials, and environmental alteration. The operation, advantages, and disadvantages of each method are covered.

After completing this chapter, the reader should be able to
- understand the function of coatings in the prevention of corrosion;
- know the basic procedures used for cathodic protection;
- know how to avoid combinations of metals or alloys that cause bimetal corrosion cells;
- have an understanding of electrolyte (environment) alteration for corrosion prevention;
- recognize specific precautions required for certain pipe materials in certain environments.

COATINGS

Coatings control corrosion by creating a barrier between a metal and its environment, the potentially corrosive electrolyte. A coating's effectiveness depends on its degree of integrity (freedom from holes, called *holidays*), its ability to bond to metal, and its ability to insulate against the flow of electrical current. A coating material must also be economically feasible.

General Properties of Coatings

Coatings were used in early attempts to control corrosion—examples include coal-tar pitch for coating cast-iron and steel pipes, and the use of various paints to preserve wood structures. Yet even with the myriad of coating materials now available, the perfect coating does not exist. Although the use of coatings dramatically reduces corrosion in water utility structures, the user must be aware of certain possible shortcomings

- factory-applied coatings may be damaged during shipping, handling, storage, or construction. It is often possible to field repair damaged coatings, but some factory-applied coatings cannot be repaired to their original quality;
- if a pinhole or holiday exists in a bonded organic coating in a corrosive environment, a concentration cell may develop that will undermine the coating and cause corrosion similar to crevice corrosion;
- if a metal with a bonded organic type coating is under cathodic protection, excessive current can cause coating disbondment.

For these reasons, it is generally accepted that bonded organic coatings in corrosive environments should be supplemented by cathodic protection. In these situations, the properties of coatings and cathodic protection systems are synergistic: coatings greatly reduce costs of the cathodic protection system, while cathodic protection substantially extends the useful life of the coating. In most instances, lower maintenance and life-cycle costs are the result.

Specific Properties of Common Coatings

The following paragraphs describe some of the commonly used coating materials. The list is not comprehensive. Other materials have been used, and still more may be developed in the future.

Asphalt. The predominant ingredient of asphalt coatings is bitumen. Most asphalt is a petroleum distillate; however, natural asphalts, such as gilsonite, are sometimes used in combination with petroleum asphalts to enhance certain physical properties.

Coal-tar epoxy and coal-tar urethane. Both coal-tar epoxy coatings and coal-tar urethane coatings are combinations of coal tar and chemicals that yield increased qualities of toughness, resistance to weathering and aging, and electrical resistance. These materials serve well as external coatings.

Coal tar (thermoplastic). The basic ingredient of coal-tar coatings is coal-tar pitch. Both hot- and cold-applied coatings are available. Coal tar is an effective coating material when properly applied with adequate thickness.

Concrete and cement mortar. Both concrete and cement mortar are used effectively for coating tanks and pipes. Concrete has the advantage that minor cracking can heal in water exposure. (For further information, review the discussion of concrete cylinder pipe in the previous chapter.)

Epoxies. A wide range of epoxy resins are available for coatings, with a wide variety of properties depending on formulation. In general, they form good, continuous coatings. They offer good resistance to alkalies, salts, oils, abrasion, weathering, and aging and have high electrical insulating ability. Adhesion is good on both metal and concrete. High-build epoxies have been especially useful in the protection of water treatment plant equipment and wells.

Fluorocarbons. Often used as thin-film coatings, fluorocarbons offer resistance to chlorine, bromine, and iodine, but in severe exposures they may be vulnerable to fluorine. They may find use in atmospheric exposures in water treatment plants where free chlorine is present. Application requires skill and experience, and cost is significant.

Glass (ceramic). Properly applied, glass coatings offer superior service in water exposures. Glass coating is a shop procedure requiring very high temperature of the metal being coated. The temperature is so great that care must be taken not to alter the metallurgical character of the metal or alloy. Glass coatings are expensive, but may be worth the cost for special applications.

Metallic coatings. Zinc, nickel, tin, and cadmium are examples of metallic coatings. With the exception of zinc and cadmium, which are sacrificial metals, their chief function is to serve as a physical barrier between the protected metal and the environment; unlike organic coatings, they do not act as a dielectric. Metal coatings are applied by electroplating, flame spraying, hot dipping, or vapor deposition. Galvanized (zinc-coated) pipes and other equipment have been used extensively in water systems, especially in service piping. If a corrosion cell develops in a galvanized pipe, the zinc, because of its position in the galvanic series, becomes anodic and tends to create cathodic (noncorrosive) conditions at the attack site. However, experience has shown that the protection afforded is short-term at best.

Polyesters. Although they are resistant to many chemicals and moisture, polyesters generally must be used with reinforcement because of their brittle nature. Adherence to metal is not good unless the metal is deeply etched or blasted. Polyester coatings are not expected to find significant application in the water supply field.

Polyether. Application requirements limit the use of chlorinated polyether in water utility service. The base resin is resistant to aging, abrasion, weather, and many chemicals. The material must be applied at elevated temperatures, then quenched in cold water. There is no solvent available that allows application at room temperature.

Polyethylene. Low- and high-density polyethylene resins are available. Although the base resin is lower in cost than most other coating materials, application procedures and service limitations may offset the higher cost of other coatings. In general, best results are obtained when the resin is applied to a metal or alloy at elevated temperatures or flame sprayed. The coating is not entirely homogeneous and, therefore, its properties of resistance are not precise. As discussed in a later section, polyethylene also is used as a wrapping material for pipe in corrosive soil environments. Its dielectric strength is excellent, as is its resistance to aging.

Polymer concretes. Polymer concretes provide high dielectric strength and resistance to penetration by chlorides. Toughness can be increased by the addition of glass fibers.

Polyvinyl chloride. Polyvinyl chloride (PVC) coatings are a versatile group of materials offering good properties of resistance and dielectric strength. They are applied by dipping, spraying, brushing, or roller coating, but usually require that the metal be blast cleaned and primed. PVC is also used for tape coatings, which offer good resistance to aging and have good dielectric strength.

CATHODIC PROTECTION

Cathodic protection is a system for reducing corrosion of a metal structure by turning the entire structure into the cathode of a galvanic or electrolytic corrosion cell. Direct electrical current, either generated by the galvanic cell or fed into the electrolytic cell from an external source, flows into the protected structure, overcoming any currents that might be created by naturally occurring corrosion cells in which the structure would be the anode. Since current does not flow from the structure to the electrolyte, corrosion cannot occur. The cathodic protection methods discussed in this section can be used for steel (dielectric or concrete-coated), cast-iron, ductile-iron, copper, and concrete cylinder pipe (reinforced, prestressed, and pretensioned).

38 EXTERNAL CORROSION

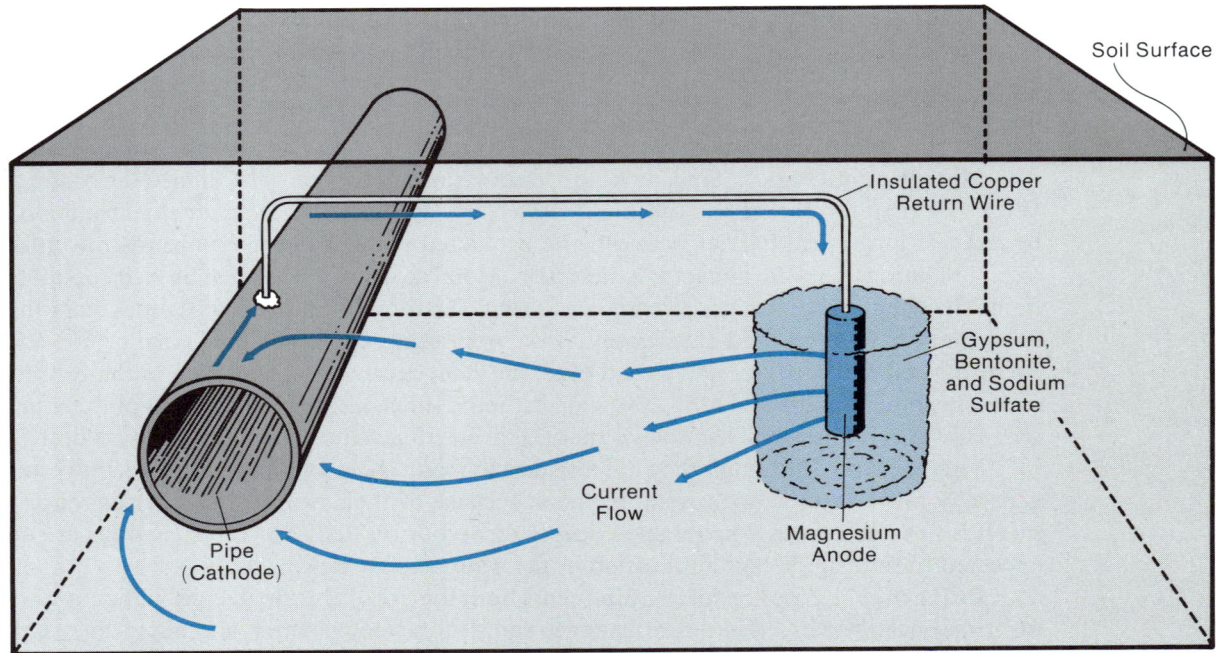

Figure 4-1 Cathodic-Protection System (Galvanic Cell Using Sacrificial Anodes)

To form the protective galvanic or electrolytic cell, the same four elements are required as for a naturally occurring corrosion cell: an anode, a cathode, a conductive electrolyte, and a return current path. The cathode is the protected structure itself, which must be electrically continuous to ensure complete protection. The conductive electrolyte is the corrosive soil environment. The anode and the return current path are added, as shown in Figure 4-1. To create a galvanic cell, anodes are selected from the galvanic series to generate the needed current. To create an electrolytic corrosion cell, an external direct current source is used. In either case, the installed anodes are bars or rods of metal that are intended to corrode; anodes may require replacement after several years of service.

The design of effective, economically feasible cathodic-protection systems is a task that should be accomplished by a qualified corrosion engineer. The examples in this chapter are not intended as a reference for design, but rather to provide an overview of procedures and equipment, as well as to show that there is a specific, dependable design system. Details such as sizing of anodes, anode material, and terminal box design must vary considerably from one installation to another, in order to achieve maximum protection and greatest economy.

CAUTION: Systems that incorporate cathodic protection require routine monitoring and maintenance to assure that proper protection is provided, and owners of these systems should be advised.

Sacrificial-Anode Systems

Operating principles and common uses. Cathodic-protection installations in which a galvanic cell is created are referred to as sacrificial-anode systems. The sacrificial anodes, also called galvanic anodes, are made of metals or alloys that are electronegative to the structure that must be protected—that is, they are nearer to the anodic end of the galvanic series. When inserted in the same soil as the structure and connected to it by a return current path, the sacrificial anode becomes the anode of a galvanic cell and corrodes, generating an electrical current. The structure becomes the cathode and is protected. Details of typical sacrificial-anode installation are shown in Figure 4-2.

PREVENTING OR REDUCING CORROSION 39

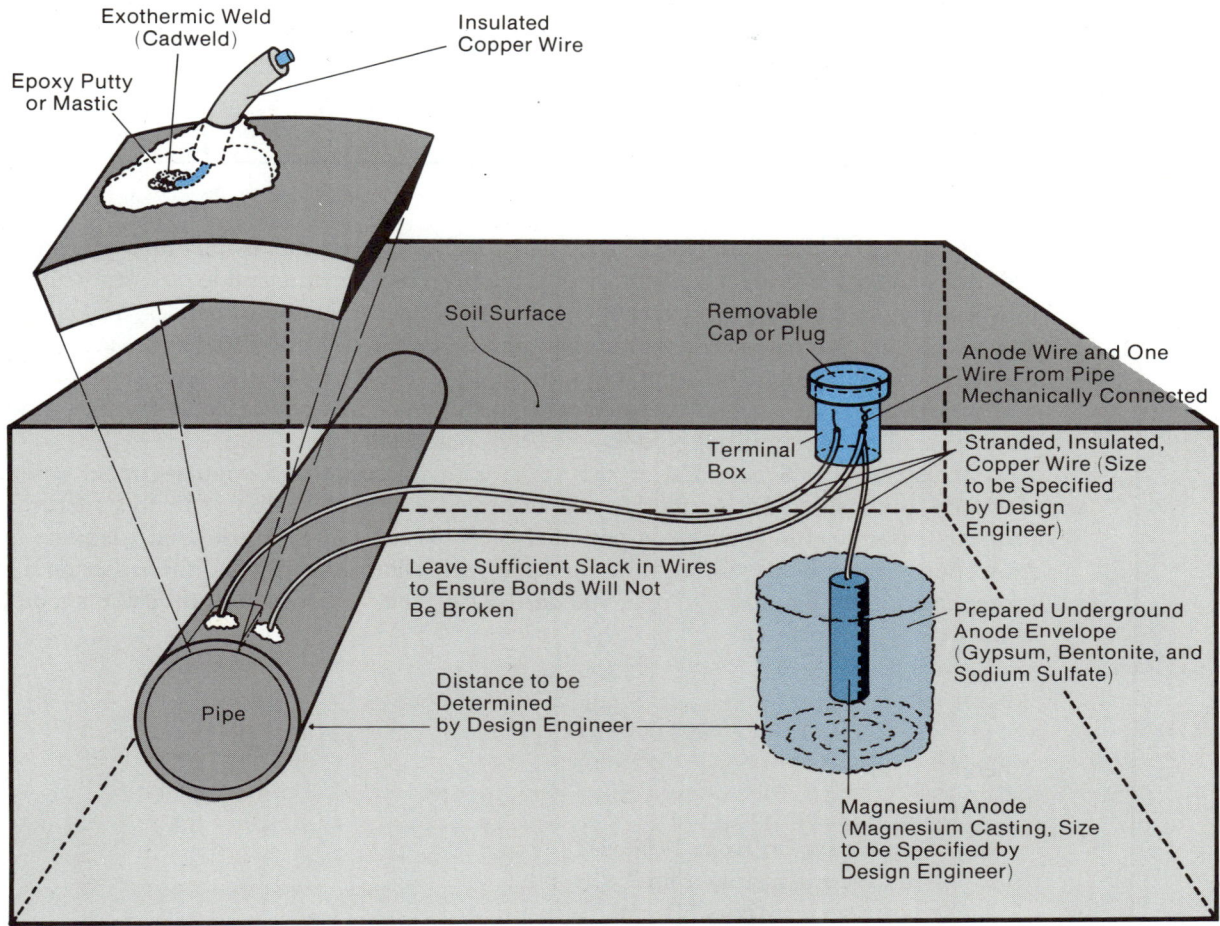

Figure 4-2 Details of a Sacrificial-Anode Installation

Sacrificial-anode protection may be used selectively in "hot spot" areas that have been located by soil-survey procedures. This requires that rubber-gasketed joints be electrically bonded through the protected area. The same protection can also be used to protect steel tanks, well casings, steel piping, and other water treatment equipment. Advantages and disadvantages of sacrificial-anode systems include the following:

Advantages
- No external power is required.
- Maintenance is minimal.
- Systems seldom cause interference with foreign structures.

Disadvantages
- Driving voltages are limited.
- Current output is limited.
- Operation is effective only in relatively low-resistivity soils.

Table 4-1 Magnesium Anode Factors

Magnesium* Anode Weight lb (kg)	Dimensions in.	(mm)	Anode Factor f	Desired P/S	Correction Factor (Magnesium) y
3 (1)	3 × 3 × 4.5	(76 × 76 × 144)	0.53	−.70	1.14
5 (2)	3 × 3 × 7.5	(76 × 76 × 191)	0.60	−.80	1.07
9 (4)	3 × 3 × 13.5	(76 × 76 × 343)	0.71	−.85	1.00
17 (8)	3 × 3 × 25.5	(76 × 76 × 648)	1.00	−.90	0.93
32 (15)	5 × 5 × 21	(127 × 127 × 533)	1.06	−1.00	0.79
50 (23)	8 diameter × 15	(203 diameter × 381)	1.09	−1.10	0.64

*These factors are based on typical anode dimensions.

Design and construction. Zinc and magnesium are metals commonly used for sacrificial anodes to protect iron and steel structures. Magnesium anodes are probably the most widely used, because they create a higher cell voltage than do zinc anodes. Magnesium anodes afford protection in soil with resistivities below 5000 Ω-cm. Zinc anodes are limited to use in soils of about 1000 Ω-cm resistivity or less. To determine the current output of a magnesium anode, the anode and correction factors must be known. Table 4-1 lists anode factors based on the required pipe-to-soil (P/S) potential.

Generally, a P/S potential of −850 mV, with reference to a copper–copper sulfate electrode, is considered to be adequate to ensure that a ferrous (iron or steel) structure is protected. For steel in concrete, a potential of −710 mV may be considered adequate for protection. Protective levels required for other metals and alloys and additional criteria of protection can be found in NACE Standard RP-01-69.* Current output of the anode is calculated by the formula

$$I = \frac{150\,000\,fy}{P} \qquad (4\text{-}1)$$

Where:

I = current output (mA)
f = anode factor, from Table 4-1
y = correction factor, from Table 4-1
P = average resistivity (Ω-cm).

For example, for a P/S potential of −0.90 V, a 17-lb magnesium anode, and an average resistivity of 1000 Ω-cm, the anode current output would be

$$I = \frac{150\,000 \times 1.0 \times 0.93}{1000} = -140 \text{ mA}$$

Anode life is estimated by the following formula:

$$L = \frac{57\,W}{I} \qquad (4\text{-}2)$$

Where:

L = life expectancy (years)
W = weight of anode (lb)
I = output current of anode (mA).

---

*National Association of Corrosion Engineers (NACE), 1440 South Creek, Houston, TX 77084.

As an example, a 17-lb magnesium anode at a current output of 140 mA would have an estimated life of

$$L = \frac{57 \times 17}{140} = 7.0 \text{ years}$$

The contact resistance of a vertically installed ground rod can be calculated as follows:

$$R_v = \frac{\rho \left(\ln \frac{8L}{d} - 1\right)}{2\pi L} \quad (4\text{-}3)$$

Where:

R_v = resistance to earth (Ω)
L = length of ground rod (cm)
d = diameter of ground rod (cm)
ρ = resistivity (Ω-cm).

Thus, it can be seen that the ground rod or chemical backfill geometry can have a great effect on the resistance. The current output is related to the resistance in accordance with Ohm's law.

$$I = E/R \quad (4\text{-}4)$$

Where:

I = current (A)
E = voltage (V)
R = resistance (Ω).

The anode factors developed by Tefankjian are the results of the foregoing computations. For example, there is a 17-lb Galvomag magnesium anode of dimensions 3 in. × 3 in. × 25½ in. packaged in chemical backfill of dimensions 6 in. × 28 in. There is also a 17-lb high-purity magnesium anode of dimensions 4 in. × 4 in. × 17 in. packaged in chemical backfill of dimensions 6½ in. × 19 in. The former will provide lower contact resistance owing to its greater length. Moreover, the former has a higher driving voltage. Thus, in accordance with Ohm's law, the former will provide more current and hence greater protective potential.

Impressed Current Systems

Operating principles and common uses. Cathodic-protection systems in which an electrolytic corrosion cell is created are called impressed-current systems, or rectifier-ground-bed systems. As shown in Figure 4-3, impressed-current systems include a rectifier unit to supply direct electrical current and a series of anodes inserted into the soil (called the ground bed). The electrical current is driven from the positive terminal of the rectifier to the ground-bed anodes, where it is driven (impressed) into the earth. The current travels through the earth and enters the protected structure. The structure carries the current to a connecting wire, which returns it to the negative terminal of the rectifier. As the current flows, the structure—acting as the cathode of the electrolytic cell—is protected from corrosion; the anodes in the ground bed corrode.

The impressed-current system is used to protect large structures or long stretches of pipeline. Advantages and disadvantages of the system include the following:

Advantages
- Large driving voltages are possible.
- Higher (virtually unlimited) current levels are available.
- The system is suitable for high-resistivity soils.
- Larger structures and longer pipelines can be protected.

Disadvantages
- Installation costs are higher than for sacrificial-anode systems.
- Maintenance and operating costs are higher than for sacrificial-anode systems.
- There is a danger of damage to other stuctures from stray currents.

Design and construction. Design of an impressed-current system requires that the corrosion-control engineer carefully evaluate the anode materials and ground-bed configuration, the current supply unit, the electrical continuity of the protected structure, the potential for stray-current corrosion of nearby structures, and a number of environmental variables.

The anodes making up the ground bed are usually of graphite, high-silicon cast iron, or scrap metal. There are four basic configurations of ground-bed installation. In the horizontally *remote* configuration, the ground bed is placed some distance from the protected structure, in order to effect a wide spread of current. In the *deep well* (vertically remote) configuration, anodes are placed in deep holes to spread current or to protect deep structures, such as well casings. In the *distributed* configuration (Figure 4-3), anodes are

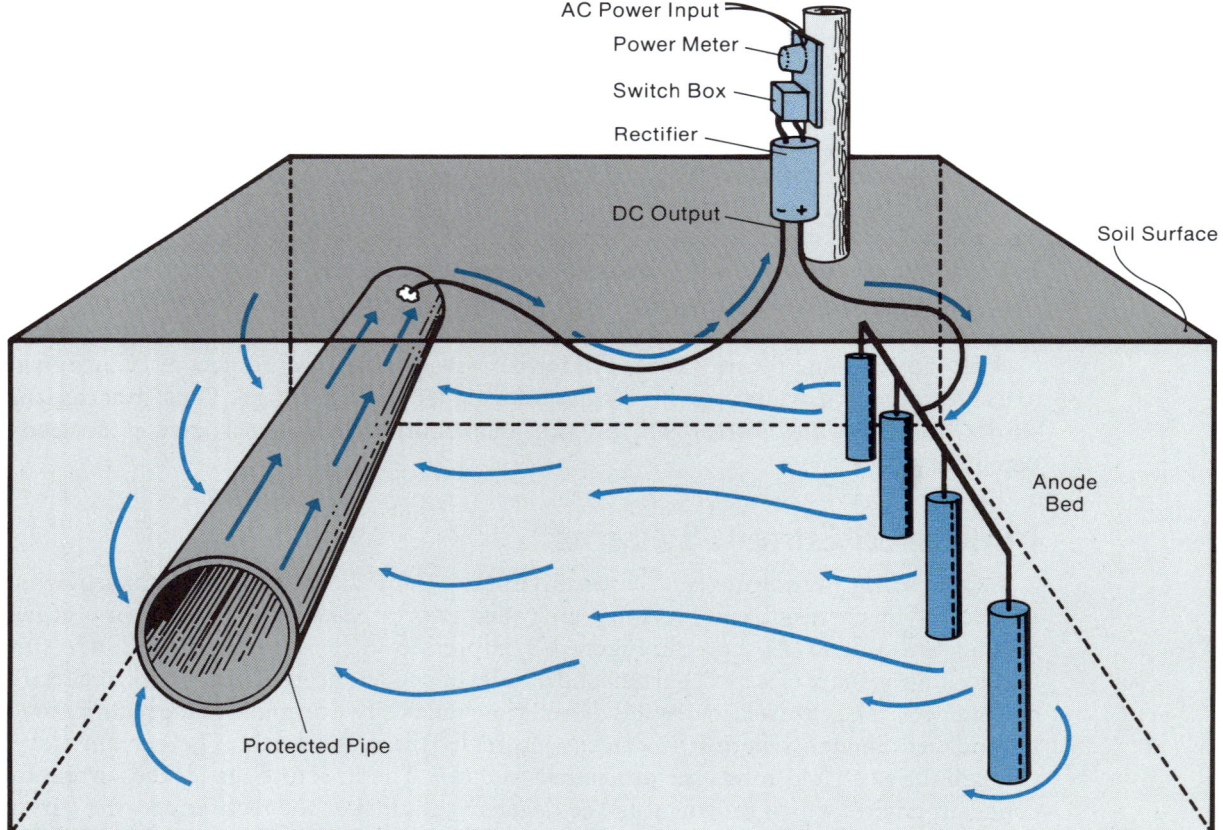

Figure 4-3 Details of an Impressed-Current System

located to protect specific structures, such as tanks, or are distributed along a pipeline; this configuration reduces interference with other underground structures and reduces the shielding effect one structure may have for another. In the *horizontal* configuration, also called the parallel configuration, a continuous anode is installed parallel to the protected pipeline. This provides good current coverage and reduces interference with other structures.

The current supply unit, as shown in Figure 4-3, includes a 110-V AC power supply, a meter, a switch box with a circuit breaker to protect the rectifier, a step-down transformer to reduce the voltage, and a rectifying element to change the alternating current to direct current. The rectifying elements are copper oxide cells, selenium cells, or silicon diodes.

The electrical continuity of the protected structure is essential, since it acts as part of the return current path. If there is a break in the electrical continuity, perhaps at a pipe joint, then the electrical driving force of the system will force the current into the soil to go around the insulated area. Corrosion will occur at the point where the current leaves the protected structure (Figure 4-4).

Metal structures that lie in the area of the protected structure but are not electrically continuous with it may also be corroded by the cathodic-protection currents (Figure 4-5). This problem, called stray-current corrosion, is a major hazard of the impressed-current system. The extent of stray-current corrosion depends on the location, size, and configuration of the foreign structure. The problem can sometimes be avoided by electrically bonding the foreign structure to the protected structure. Refer to Chapter 3 for a discussion of stray-current monitoring, which must be part of any cathodic-protection system.

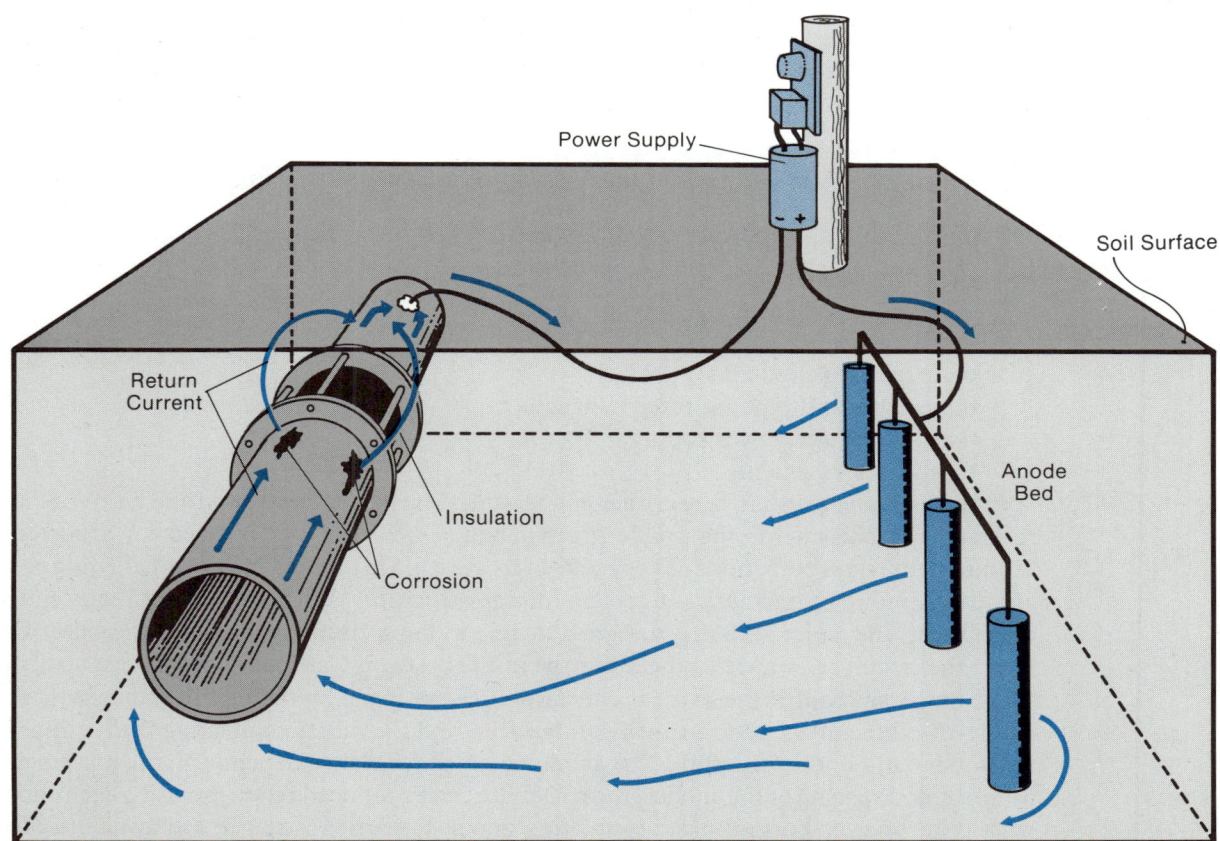

Figure 4-4 Corrosion Caused by Electrical Discontinuity in a Cathodically Protected Pipeline

44 EXTERNAL CORROSION

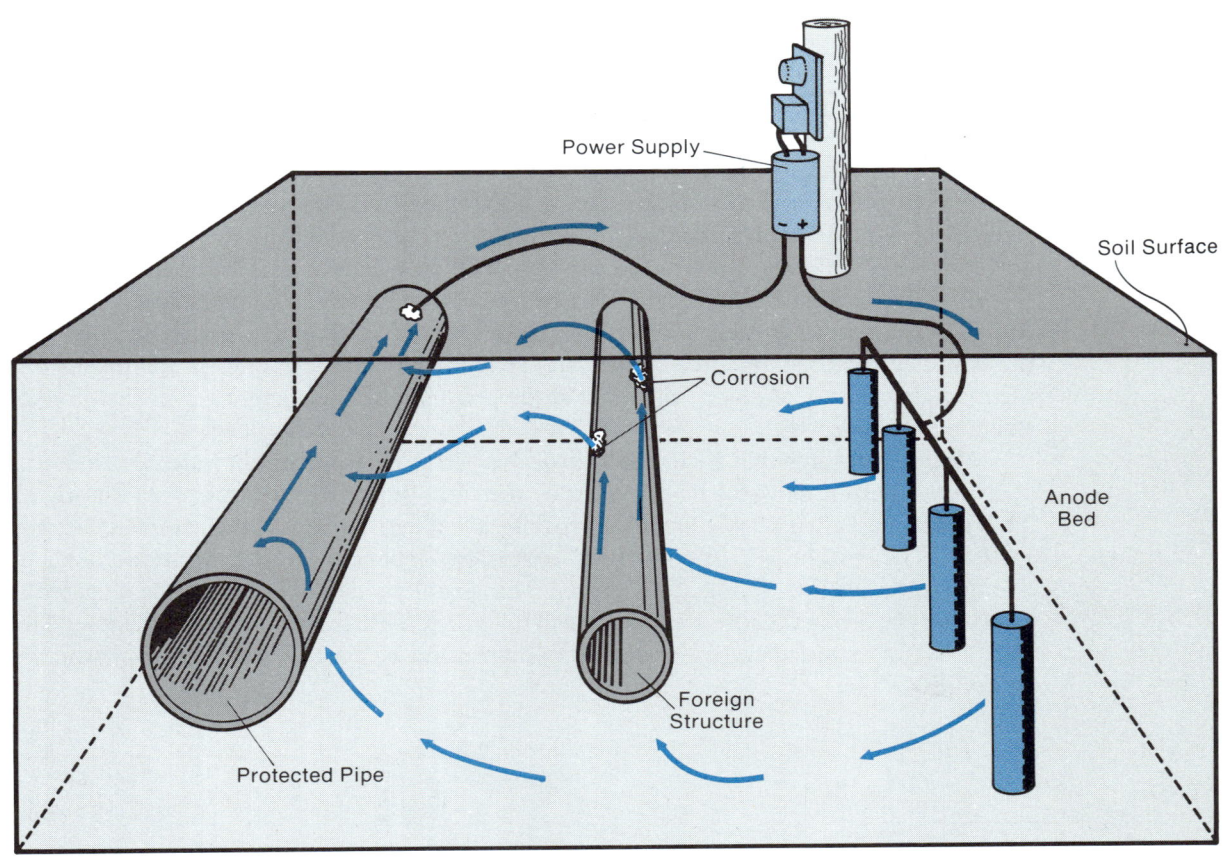

Figure 4-5 Corrosion of a Metal Structure in the Vicinity of a Cathodically Protected Structure

Environmental variables to be considered when designing an impressed-current cathodic-protection system include
- details of the structure to be protected;
- average soil resistivity;
- current requirements;
- locations and types of foreign structures;
- availability of power supply;
- property easements.

In designing a system to serve a pipeline, soil-survey data is reviewed and an area of low resistivity is chosen for the anode ground bed. Once the area is selected, a current-requirement test is performed. The results of this test aid in selecting the rectifier equipment and determining the potential effects on foreign structures. With the current requirement established, the entire system is designed to have as low a circuit resistance as feasible. The lower the circuit resistance is, the lower the initial cost and continuing power costs will be.

Significant contributions to circuit resistance are made by the rectifier, the earth, the anode-to-earth and structure-to-earth interfaces, and the cables connecting the rectifier to the anodes and to the structure. The anode-to-earth resistance depends on the placement, number, and spacing of the anodes; the resistance can be reduced by surrounding each anode with coke breeze. Formulae for estimating anode-to-earth resistance are available. The number of anodes can also be determined by a formula if anode-to-earth resistance, soil resistivity, anode spacing, anode length, and anode radius have been determined.

POLYETHYLENE ENCASEMENT OF GRAY AND DUCTILE CAST-IRON PIPE

Ductile cast-iron pipe is manufactured in 18- and 20-ft (5.5-m and 6-m) lengths, and all modern joints use rubber gaskets. Because of this, a ductile cast-iron pipeline is a series of electrically segmented units. Experience has shown that the joints retain resistance even after years of burial. Older gray cast-iron pipe was installed using other types of joints, such as caulked lead, sulfur compound, and cement, but the resistance of those joints is high enough that such pipelines are also considered electrically noncontinuous.

After years of study using adherent coating, trench improvement, and cathodic protection, the system of wrapping with 8-mil (0.008-in.) thick, loose polyethylene encasement was researched and resulted in the development of ANSI/AWWA C105/A21.5, Standard for Polyethylene Encasement for Ductile-Iron Piping for Water and Other Liquids. There are three methods of installing the encasement, as shown in Figure 4-6. Method A, illustrated in Figure 4-6, is the most widely used.

Method A: One length of polyethylene tube for each length of pipe, overlapped at joint.

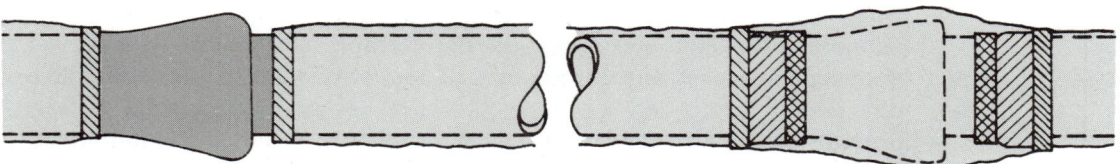

Method B: Separate pieces of polyethylene tube for barrel of pipe and for joints. Tube over joints overlaps tube encasing barrel.

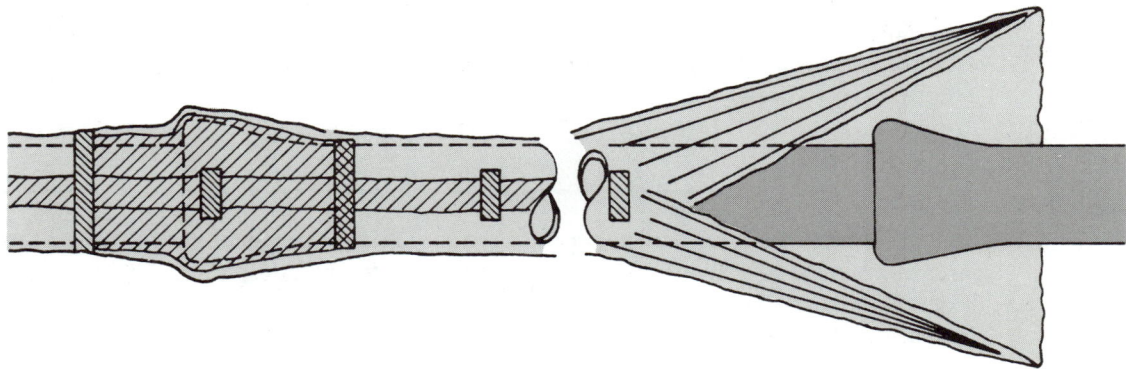

Method C: Pipeline completely wrapped with flat polyethylene sheet.

Figure 4-6   Three Methods for Polyethylene Encasement of Cast-Iron Pipelines

Polyethylene encasement is not a coating, although it offers some of the qualities of a coating, such as dielectric strength. It is mainly an environmental improvement. First, it reduces the environment to a very thin space between the pipe and the loose wrap; second, it excludes direct exposure to corrosive soil. It does allow the entrance of groundwater into the annular space between the pipe and wrap; however, the corrosive characteristics of the water are soon depleted by the action of initial corrosion, usually oxidation. Extensive use of polyethylene encasement has demonstrated its effectiveness for gray and ductile cast-iron pipe in the following respects:

- it provides a uniform environment for the pipe, which eliminates local galvanic corrosion cells;
- it offers good electrical shielding, which resists the assimilation of stray direct current;
- because it is installed on the pipe at the time of pipeline construction, it is less likely to be damaged than factory applied coatings;
- pinholes and minor damage to the loose wrapping material do not diminish its protective ability. Rips, punctures, or other significant damage to the polyethylene film should be repaired;
- initial cost of material and installation are very low, and there are no maintenance costs.

MATERIALS SELECTION

Earlier chapters discussed materials and combinations of materials that are vulnerable to corrosion under certain conditions. Based on that discussion, the following rules of thumb may be helpful in reducing the corrosion problems often associated with water utility equipment:

- avoid the use of combinations of metals that lie far apart in the galvanic series;
- when small metal units are in direct contact with larger masses of metal, they should be more noble than the larger mass. For example, mechanical bolts of steel alloyed with copper, nickel, and chromium will not corrode when used in ductile cast-iron pipe joints;
- provide the best available coatings and paints in enclosed, humid environments within the treatment plant and pumping stations, especially those where chlorine-bearing atmosphere occurs;
- avoid extensive reliance on sacrificial coatings, such as galvanizing (zinc) on permanent structures;
- recognize the passivating effect of structural constituents. For example, carbon in gray and ductile cast iron, concrete on steel, and noble metals as constituents of steel alloys will all increase corrosion resistance;
- watch for sites of stress, fatigue, fretting, and crevice corrosion, and specify metals that are most resistant to these problems.

TRENCH IMPROVEMENT

Trench improvement of corrosive soils to reduce their corrosive tendencies is, for the most part, impractical. Extensive studies have been made of bedding and backfilling around underground pipe with sand, limestone, limestone screenings (fines), dolomite, sand plus 10 percent cement, and select soil. Although some reduction in corrosion incidence is realized with most of these treatments, it has been observed that, in many instances, the substitute fill material eventually takes on the characteristics of the surrounding soil. The long-term result can be inadequate protection.

PROTECTIVE METHODS FOR SPECIFIC PIPE MATERIALS

Experience has shown that, for any given pipe materials, certain protective measures or combinations of protective measures are generally most effective and most economically feasible. This section briefly considers several of the pipe materials commonly used in water utility systems, with attention to the protective measures that have often been found appropriate.

Steel Pipe

Coatings for corrosion control of steel pipe are extremely effective when properly used. They are considered to be the primary line of defense against corrosion of steel pipeline systems. The American Water Works Association has developed standards for widely used protective coatings that have provided years of effective corrosion control for buried steel pipelines. The coatings currently covered include coal-tar enamel, shop-applied cement-mortar coating, cold-applied tapes, coal-tar epoxy coatings, fusion-bonded epoxy coatings, and tape coating systems.

Under certain environmental conditions, cathodic protection of steel pipe may be appropriate, usually in addition to coatings. If a steel pipeline is to be cathodically protected, or if the pipeline being installed may require cathodic protection at some future time, the joints must be bonded to make the line electrically continuous. It is generally best to bond all joints at the time of installation, because the cost later will be many times greater. Field-welded lines are inherently bonded. In addition to bonding, test leads should be connected to the pipeline at appropriate intervals to permit monitoring of the activity of electrical currents within the pipeline, whether it is under cathodic protection or not.

Copper Pipe

Copper exposed to aggressive conditions should be isolated from the environment with an inert moisture barrier, a wrapping of insulating tape, a coating of an asphaltum paint, or with some other approved material. Cathodic protection may also be necessary if copper pipe is coated.

Concrete Cylinder Pipe

Where concrete cylinder pipe is installed in acidic soils (pH of 5 or less), and where considerable groundwater movement is anticipated, the pipe may require one or more of the following precautions:

- compacted and relatively impermeable backfill material (for example, clay) completely surrounding the pipe will effectively inhibit acid groundwater replenishment in the immediate pipe zone;
- calcareous backfill material will neutralize the acid and maintain an alkaline or neutral pH zone around the pipe;
- coating the pipe exterior with a barrier coating, such as a cut back asphalt or coal tar, is an alternative protective measure.

Concrete cylinder pipe installed in sulfate soil areas has demonstrated excellent sulfate resistance. Field experience, as well as research by both the Portland Cement Association and the US Bureau of Reclamation, has established that dramatic increases in sulfate resistance are achieved by increasing cement factors. Criteria involving the reduction of allowable tricalcium aluminate content of the cement are found, in tabular form, in many engineering handbooks relating to concrete. Such criteria are intended for application to conventional, field-placed concrete structures, and are considered to be conservative for buried concrete cylinder pipe.

When concrete cylinder pipe will be exposed to the atmosphere, additional protective measures are desirable. The external surface of the pipe can be sealed to prevent the admittance of moisture. The surface coating should be a light color to reflect sunlight. Materials such as epoxies and vinyls are recommended. Scheduled inspections and maintenance touch-up of the protective treatment are considered good practice. If an exposed line is installed in a salt-water spray zone or in a freezing climate, a second layer of cement-mortar coating should be applied to the pipe prior to the paint coating. If deterioration occurs to the outer coating layer, the inner one remains intact to protect the underlying steel structure. The outer coating can be restored to its original sound condition without permanent damage to the structure.

Gray and Ductile Cast-Iron Pipe

The polyethylene encasement system for the protection of gray and ductile cast-iron pipe is discussed in an earlier section of this chapter.

Glossary

(Glossary terms are defined at the back of the book.)

Cathodic protection
Coating
Environmental alteration
Holiday
Hot spot
Impressed-current system

Mitigate
Polyethylene encasement
Rectifier-ground-bed system
Sacrificial anode
Sacrificial-anode system
Stray current

Glossary

aggressive Corrosive.

anode That part of a corrosion cell that is discharging current into the electrolyte and is corroding. Electrochemical oxidation occurs at this electrode.

backflow The entrance of water or other liquid from any but the normal source of a potable water supply system.

bacteriological corrosion Corrosion that results from the by-products of sulfate-reducing bacteria in media of very low or no oxygen content.

bimetal couple A type of corrosion where two different metals or alloys are in contact with each other in a common media.

cathode That part of a corrosion cell that is receiving current and is protected. Electrochemical reduction occurs at this electrode.

cathodic protection Reduction of corrosion by making the metal a cathode. This is done by causing direct current to enter the structure's surface at all locations.

coating An application of an adherent or mechanically bonded material between metallic structures and their environment (electrolyte).

concentration cell A corrosion cell involving two identical electrode materials, with corrosion resulting from mechanical, physical, or chemical differences of the environments adjacent to the two electrodes.

conductivity The ease with which an electrical circuit allows current to flow. Conductivity, measured in mhos or Siemens, is the reciprocal of resistivity.

copper–copper sulfate electrode (reference electrode, or CSE) A copper rod partially immersed in a copper sulfate solution inside a cylindrical tube with a porous bottom, creating the cathodic half of a galvanic cell. The copper sulfate provides excellent soil contact, much better than would the bare copper rod. A reference electrode made with this cell is used in making field measurements of pipe-to-soil potentials.

corrosion cell The arrangement of an anode and a cathode in contact with a common electrolyte in such a manner that current discharges from the anode into the electrolyte. Corrosion cells are either galvanic, which generate electrical current, or electrolytic, which are driven by an outside electric current.

corrosion control The reduction in the rate of deterioration and/or the total elimination of the environmental impact on water utility systems.

corrosion (general) Deterioration of a material by reaction with its environment.

corrosion (metallic) Deterioration of a metal by reaction with its environment. This is nature's way of returning refined metals to their natural state. Rust is iron oxide—iron ore also is iron oxide.

CSE *See* copper–copper sulfate electrode.

current The movement of electricity through a circuit, measured in amperes (A). Conventional electrical current flow is said to move through an external circuit from the positive terminal (cathode) to the negative terminal (anode) of a galvanic cell; electrons actually move in the opposite direction, from negative to positive. Electrical current is analogous to flow rate in a water pipe.

current span The amount of current moving through a pipe, calculated using the measurement of the difference in voltage between two locations along the pipeline.

dezincification Selective removal of zinc from an alloy, such as brass.

electrode One of two pieces of metal that are immersed in an electrolyte to form a corrosion cell. The corroding electrode, from which electrical current enters the electrolyte, is called the anode. The protected electrode is called the cathode.

electrolyte The ionically conductive media in which the anode and cathode of a corrosion cell are immersed.

electrolytic corrosion cell A cell in which an external direct current generates the corrosion.

electron (e^-) A constituent of an atom with a negative charge. Electrons flow in the direction opposite to current flow.

electronic path The movement of electrons through a complete electrical circuit.

environment The surrounding materials and conditions that influence the water utility system.

environmental alteration The selection of an electrolyte (backfill material) to eliminate or reduce the rate of corrosion. Also referred to as trench improvement.

Faraday's law The amount of any substance dissolved or deposited in electrolysis is proportional to the total electric charge passed.

fatigue corrosion *See* stress corrosion.

fretting corrosion *See* stress corrosion.

galvanic corrosion cell A cell that generates direct electrical current.

galvanic series A list of metals and alloys arranged according to their relative corrosion potentials in a given electrolyte.

ground-voltage gradients A record of the voltage differences between each two of a series of locations along the ground surface.

holiday A void in a coating that will allow the passage of electrical current.

hot spot An area of soil found by survey and analysis or experience to be more corrosive than surrounding soil.

impingement corrosion Localized attack caused by turbulence, cavitation, or other erosion that breaks through corrosion scale.

impressed-current system A cathodic protection system utilizing an outside source of power, converting it to direct current, and injecting it into the soil through an anode bed. Also called a rectifier-ground-bed system.

ion One of the electrically charged particles produced by the disassociation (breakup) of a chemical compound.

mitigate To moderate or make less severe.

Ohm's law The current I flowing in a circuit is equal to the voltage E divided by the resistance R. Thus, $I = E/R$, where I is in amperes, E in volts, and R in ohms.

oxidation–reduction potential (redox potential) The electrical potential (in millivolts) between platinum and reference electrodes inserted in soil, measured with a pH meter. A low potential indicates low soil aeration.

passivation A condition of metal or alloy, usually at its surface, causing it to behave in a more noble manner.

pH The hydrogen ion activity of a media. pH ranges from 0 to 14. A pH of less than 7 is acidic, 7 is neutral, and greater than 7 is alkaline.

pH meter An instrument used for the electronic determination of pH. pH can also be measured with colorimetric or other analytical chemical methods.

pipe-to-soil potential Any two connected metals develop an electrical potential (driving force in volts) when in contact with a common media, such as soil. Traditionally, pipe-to-soil potential is the potential measured between an underground metal, such as pipe, and a copper–copper sulfate electrode.

pitting corrosion Highly localized corrosion that causes penetration into the metal at a few spots.

polarization Retardation of the corrosion process by the buildup of protective layers on an electrode, usually the cathode or anode.

polar solvent A solvent compound whose molecules are polarized. That is, the electrostatic charge on one side of the molecule is relatively positive and, consequently, relatively negative on the other, due to unequal sharing of electrons in the covalent bonding between the atoms. Polar solvents separate the structural units of the solvate by surrounding each ion with a cluster of solvent molecules held to the positive or negative ion by the oppositely charged end of the solvent molecule. Water (H_2O) and methanol (CH_3OH) are examples of highly polar solvents.

polyethylene encasement An 8-mil (0.008-in.) polyethylene film placed around gray or ductile cast-iron pipe to prevent corrosion.

potential The force available to drive an electrical current through a circuit, measured in volts (V). It is analogous to pressure (head) in a water pipe.

rectifier-ground-bed system *See* impressed-current system.

resistance The tendency of an electrical circuit to retard the flow of current. It is measured in ohms (Ω), and is analogous to friction factor in a water pipe.

resistivity A measure of the effective resistance of a media, such as soil, over a given distance. It is measured in ohm-centimetres (Ω-cm).

resistivity meter An instrument that utilizes batteries and is used to measure the average resistance to current flow in a media.

return current path The metallic connection between the anode and cathode of an electrochemical cell.

return on investment A method to determine whether any proposed course of action will prove to be economical, as compared to other possible alternatives.

sacrificial anode An anode of metal less noble than a metal structure to be cathodically protected. It is sacrificed by corrosion to create cathodic conditions at the protected structure.

sacrificial-anode system A cathodic protection system in which the driving voltage for the protective current is generated by a galvanic corrosion cell, with the protected structure being the cathode.

single probe A probe, usually 4–4½ ft (1.2–1.4 m) in length, used to measure resistivity soil around the tip or point of the probe.

soil box A device used to determine resistivity of a confined volume of soil. The box is manufactured by several companies; however, one can be constructed conveniently. It measures 4 cm × 4 cm × 4 cm and two of the opposite sides are of stainless steel.

soil corrosion Corrosion of underground materials resulting from soil conditions.

soil resistivity An indication of difficulty with which a soil conducts electrical current. The average electrical resistance of a volume of soil.

stray current Direct current traveling through the earth around an existing underground structure. If stray current enters and leaves that structure, corrosion will occur.

stress corrosion Corrosion that acts on metal at points of tensile stress, working, or vibration wear.

sulfides A chemical compound of sulfur with another element. A result of the life process of sulfate-reducing bacteria.

voltage *See* potential.

References

Basic Science Concepts and Applications. AWWA, Denver, Colo. (1980, 1984).

Concrete Pressure Pipe. AWWA Manual M9. AWWA, Denver, Colo. (1979).

PVC Pipe. AWWA Manual M23. AWWA, Denver, Colo. (1980).

Steel Pipe—A Guide for Design and Installation. AWWA Manual M11. AWWA, Denver, Colo. (1985).

American National Standard for Polyethylene Encasement for Ductile-Iron Piping for Water and Other Liquids. ANSI/AWWA C105/A21.5-82. AWWA, Denver, Colo. (1982).

Standard for Coal-Tar Protective Coatings and Linings for Steel Water Pipelines—Enamel and Tape—Hot-Applied. AWWA C203-78. AWWA, Denver, Colo. (1978).

Standard for Chlorinated Rubber-Alkyd Paint System for the Exterior of Aboveground Steel Water Piping (withdrawn). AWWA C204-75. AWWA, Denver, Colo. (1975).

Standard for Cement-Mortar Protective Lining and Coating for Steel Water Pipe—4 in. and Larger—Shop Applied. AWWA C205-85. AWWA, Denver, Colo. (1985).

Standard for Cold-Applied Tape Coatings for the Exterior of Special Sections, Connections, and Fittings for Steel Water Pipelines. AWWA C209-84. AWWA, Denver, Colo. (1984).

Standard for Liquid Epoxy Coating Systems for the Interior and Exterior of Steel Water Pipelines. AWWA C210-84. AWWA, Denver, Colo. (1984).

Standard for Fusion-Bonded Epoxy Coating for the Interior and Exterior of Steel Water Pipelines. AWWA C213-85. AWWA, Denver, Colo. (1985).

Standard for Tape Coating Systems for the Exterior of Steel Water Pipelines. AWWA C214-83. AWWA, Denver, Colo. (1983).

Standard for Reinforced Concrete Pressure Pipe, Steel-Cylinder Type, for Water and Other Liquids. AWWA C300-82. AWWA, Denver, Colo. (1982).

Standard For Prestressed Concrete Pressure Pipe, Steel-Cylinder Type, for Water and Other Liquids. AWWA C301-84. AWWA, Denver, Colo. (1984).

Standard for Reinforced Concrete Pressure Pipe—Steel Cylinder Type, Pretensioned, for Water and Other Liquids. AWWA C303-78. AWWA, Denver, Colo. (1978)

Standard Practice for the Selection of Asbestos–Cement Distribution Pipe, 4 in. Through 16 in. (100 mm Through 400 mm) for Water and Other Liquids. AWWA C401-83. AWWA, Denver, Colo. (1983).

Standard Practice for the Selection of Asbestos–Cement Transmission and Feeder Main Pipe, Sizes 18 In. Through 42 In. (450 mm Through 1050 mm). AWWA C403-84. AWWA, Denver, Colo. (1984).

Index

Acidic soils, 29
Aggressiveness, 2
Anode, 8
Asbestos-cement (AC) pipe, 29-30
Asphalt, 36-37
Atmospheric corrosion, 19, 29

Backflow, 4
Bacteriological corrosion, 18
Bimetal couple, 2, 15

Cathode, 8
Cathodic interference, 29
Cathodic protection, 5-6, 8-9, 37-38
Cavitation, 17-18
Cell voltage, 13-14
Cement mortar, 36
Chlorides, 26, 29
Coal tar, 36
Coatings, 35
   properties, 36-37
Concentration cell, 17
Concrete, 36
Concrete cylinder pipe, 26, 29
   protective methods, 47-48
Conductivity, 21
Conventional current flow, 11
Copper pipe, 24-26
   protective methods, 47
Copper-copper sulfate electrode, 32
Corrosion
   conditions favoring, 2, 4
   definition, 2
   implications, 4
Corrosion cell, 9
Corrosion chemistry, 12-19
   See also Electrochemistry
Corrosion control
   economics, 4-5
   materials selection, 46
   overview, 1, 35
   responsibility for, 5-6
Corrosion rate, 13-14
Current, 11-12
Current flow, 11, 33
Current span, 31, 33

Deep well configuration, 42
Dezincification, 19
Distributed configuration, 42-43
Ductile cast-iron pipe
   See Gray/ductile cast-iron pipe

Electrical current flow, 11
Electrochemistry
   electrolytic corrosion, 12
   galvanic corrosion, 7-9, 11
   See also Corrosion chemistry
Electrode, 7-8
Electrolyte, 8, 20
   nonuniform, 9, 11
Electrolytic corrosion, 12
Electrolytic corrosion cell, 12
Electron, 8
Electronic path, 8
Environmental contamination, 2
Environmental stress cracking, 31
Epoxies, 36

Faraday's law, 14
Fatigue corrosion, 19
Fluorocarbons, 36
Four-pin system, 21
Fretting corrosion, 19

Galvanic anode, 38
Galvanic corrosion
   chemistry, 8-9
   current flow, 11
   nonuniform electrolytes, 9, 11
   overview, 7-8
Galvanic corrosion cell, 8, 15
Galvanic couples, 15
Galvanic series, 15-16
Glass, 37
Gray/ductile cast-iron pipe, 21, 24
   polyethylene encasement, 45-46
   protective methods, 48
Ground bed, 41
Ground-voltage gradients, 31, 33-34

Holiday, 35
Horizontal configuration, 43
Horizontally remote configuration, 42
Hot spot, 39
Hydrogarnets, 29

Impingement corrosion, 17-18
Impressed-current system, 41-44
Ion, 8
IR-drop, 31-33

Metal dissimilarities, 2
Metallic coatings, 37
Moisture, 24

Ohm's law, 14
Oxidation-reduction potential, 24

Parallel configuration, 43
Passivation, 17
pH, 24
Pipe-to-soil potential, 31–33
Pitting corrosion, 17
Polarization, 16
Polybutylene (PB), 30
Polyesters, 37
Polyethylene (PE), 30, 37
Polyethylene encasement, 45–46
Polymer concretes, 37
Polyvinyl chloride (PVC), 31, 37
Potential, 13

Quad box, 21

Rectifier-ground-bed system, 41
Redox potential
    See Oxidation-reduction potential
Reference electrode, 31–32
Resistance, 15
Resistivity, 21
Return current path, 8
Return on investment (ROI), 4–5

Sacrificial anode, 38
Sacrificial-anode system, 38–41
Single probe, 21
Soil box, 21
Soil corrosion, 2, 18
Soil resistivity, 21, 24
Soil variances, 2
Solvation, 30–31
Steel pipe, 24
    protective methods, 47
Stray current, 12, 20, 31–34, 43
Stress corrosion, 19
Sulfate soils, 29
Sulfides, 24

Thermoplastic pipe, 30–31
Trench improvement, 46

Voltage, 13–14